Starting SCIENCE

BOOK THREE

TONY PARTRIDGE

Formerly Head of Science
Severn Vale School

Oxford University Press

Oxford University Press, Walton Street, Oxford OX2 6DP
Oxford New York
Athens Auckland Bangkok Bombay
Calcutta Cape Town Dar es Salaam Delhi
Florence Hong Kong Istanbul Karachi
Kuala Lumpur Madras Madrid Melbourne
Mexico City Nairobi Paris Singapore
Taipei Tokyo Toronto

and associated campanies in
Berlin Ibadan

Oxford is a trade mark of Oxford University Press

First published 1992
Reprinted 1993 (twice), 1994

ISBN 0 19 914374 9
Printed in Italy by G. Canale & C. S.p.A., Borgaro T.se, Turin

Contents

Contents

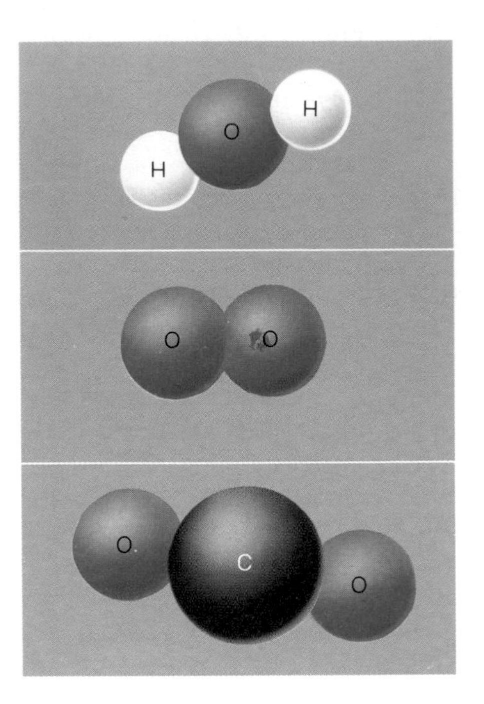

H
O
H
O
O
O
C
O

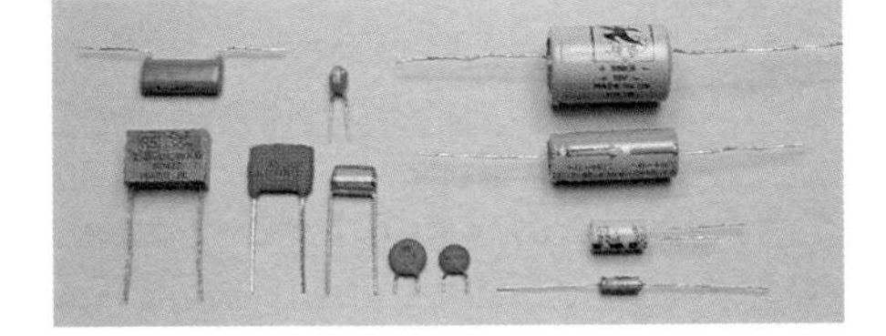

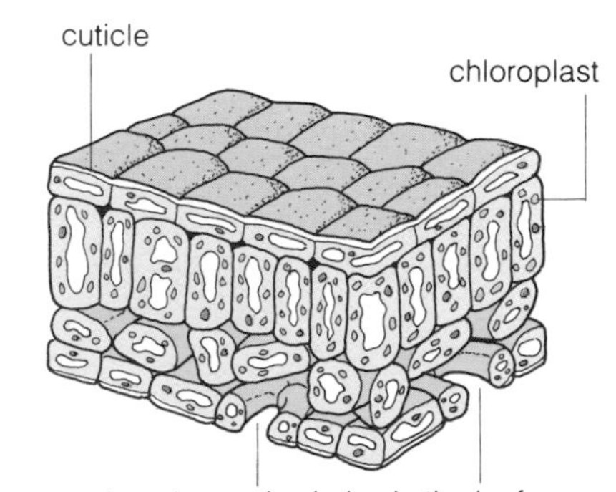

cuticle
chloroplast
stomata are tiny holes in the leaf. Oxygen and carbon dioxide pass through these holes.

Contents

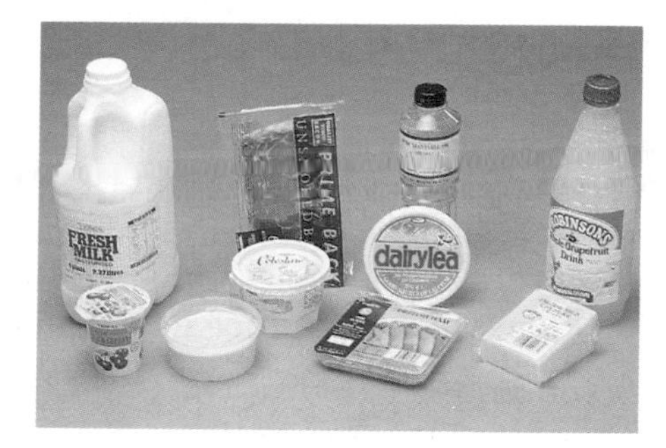

About this book . . .

Starting Science 3 follows on from **Starting Science 1** and **2**. Some sections, such as *Materials* and *Energy and environment*, build on ideas from those books. Other sections, such as *Electronics* and *The Earth in space*, introduce new subjects.

As you know, practical work and investigations are central to science. But reading is important too. You can use this book to support your practical activities in a number of ways:

- to introduce ideas at the beginning of a topic
- to help explain the results of your practical work
- to check your understanding of scientific ideas
- to help you to make links between the science you do and the way it can be put into practice
- to help you to use your science to explain everyday things.

However you use the book, I hope that you find it helpful.

Tony Partridge

And how to use it . . .

Starting Science is made up of units. Most units contain three pages.

Starting off is the first page. In it, you will learn a new piece of science. You should begin with this page. Otherwise, the other pages won't make any sense.

Going further is the second page. It follows on from what you learned in **Starting off**.

For the enthusiast, the third page, takes you even further. The material on it is usually more difficult.

When you start to work on a page, you should first read everything thoroughly - including *Did You Know*. You should also look carefully at any diagrams. Then you can answer the questions. Some questions end with a triangle sign (▲). This tells you that the answer to the question is written somewhere on the page. Some questions begin **Try to find out**. You will usually have to look through other books - like encyclopaedias - for the answers to these. To answer the other questions, you will have to use what you have learned on the page, and a bit of brain power! Using your brain is all part of **Starting Science**!

The Earth in space

16

Stonehenge, built about 4000 years ago, was a complex astronomical computer. It could be used to predict when eclipses would occur.

The radio telescope at Jodrell Bank, in Cheshire, can detect radio waves from objects far out in space. Scientists interpret this information and are putting foward new theories about the Universe.

Astronomy, ancient and modern

The study of the stars and planets has always been important to human beings.

In the Stone Age, before calendars were developed, the movement of the stars was used to plan the sowing times for crops.

Five thousand years ago, priests in the Middle East were making astronomical measurements. They could predict eclipses, and used their information for ceremonies and for astrology.

In the 1700s, astronomers collected accurate information on the positions of stars and planets so that ocean navigation could become much safer.

Nowadays, astronomers continue to try to learn about the nature of the Universe. Attempts to explore space have given us new knowledge, and have led to the development of new materials.

16.1 The Sun and stars seem to move

Starting off

Shadow stick

These students are out in the sunlight. They have put a vertical stick about a metre long in the ground. They are marking the end of the shadow with a stone. In an hour they will come back and mark the new position of the shadow.

Motion of the stars

The pole star (**Polaris**) appears almost directly overhead at the North Pole. Its position in the sky never seems to change. All the other stars appear to move round the pole star each night. The photograph shows the paths of many stars during one night.

So, each *day* the **Sun** appears to move across the sky. And each *night* the **stars** appear to move across the sky. Could these movements be caused by the same thing? Let's try to find out.

On a roundabout

Imagine that you are standing on a roundabout, spinning anticlockwise. As you spin round, your surroundings seem to flash past you from left to right. But really, the surroundings aren't moving. They *appear* to move as a result of your spin.

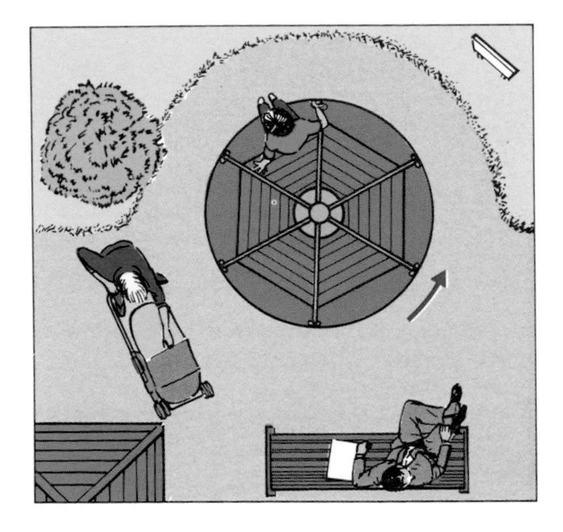

The way the Sun and the stars appear to move across the sky is caused by the spinning of the Earth. The Earth spins once in about 24 hours. Because of the Earth's spin, the Sun and stars *appear* to be overhead at different times in different places.

1 Which way would the shadow of the stick have moved?
2 What does this show about the direction in which the Sun seems to move?
3 In your book, copy the top view of you on the roundabout. Describe how the surrounding objects would appear to you as you moved through one complete turn. ▲
4 If it was a sunny day, how would the Sun appear to move as you spin through one complete turn? ▲
5 Explain how the Earth's spin makes the Sun and the stars appear to move. ▲
6 In the star photograph, estimate how long the camera's shutter was open.

Did you know ?

- From early times people have used the apparent movement of the Sun to tell the time. Sundials, though, rarely tell exactly the same time as 'clock' time. At different times of the year they can be up to a quarter of an hour 'fast' or 'slow'.

16.1 Flat or spherical?

Going further

On Sunday, September 7th 1522, a small wooden ship limped into a harbour in southern Spain. Her top-mast was damaged and she was leaking badly. Yet the **Victoria** and her crew were treated like heroes.

They had made the first recorded voyage right round the world. Many friends were surprised to see them at all. "We thought you would have sailed off the edge of the Earth!"

From our point of view it may be obvious that the world is spherical. We can see photographs of the Earth taken from space, but before these were available, what was the evidence? For more than 5000 years people have had ideas, or made **hypotheses**, about the shape of the Earth and its position in space.

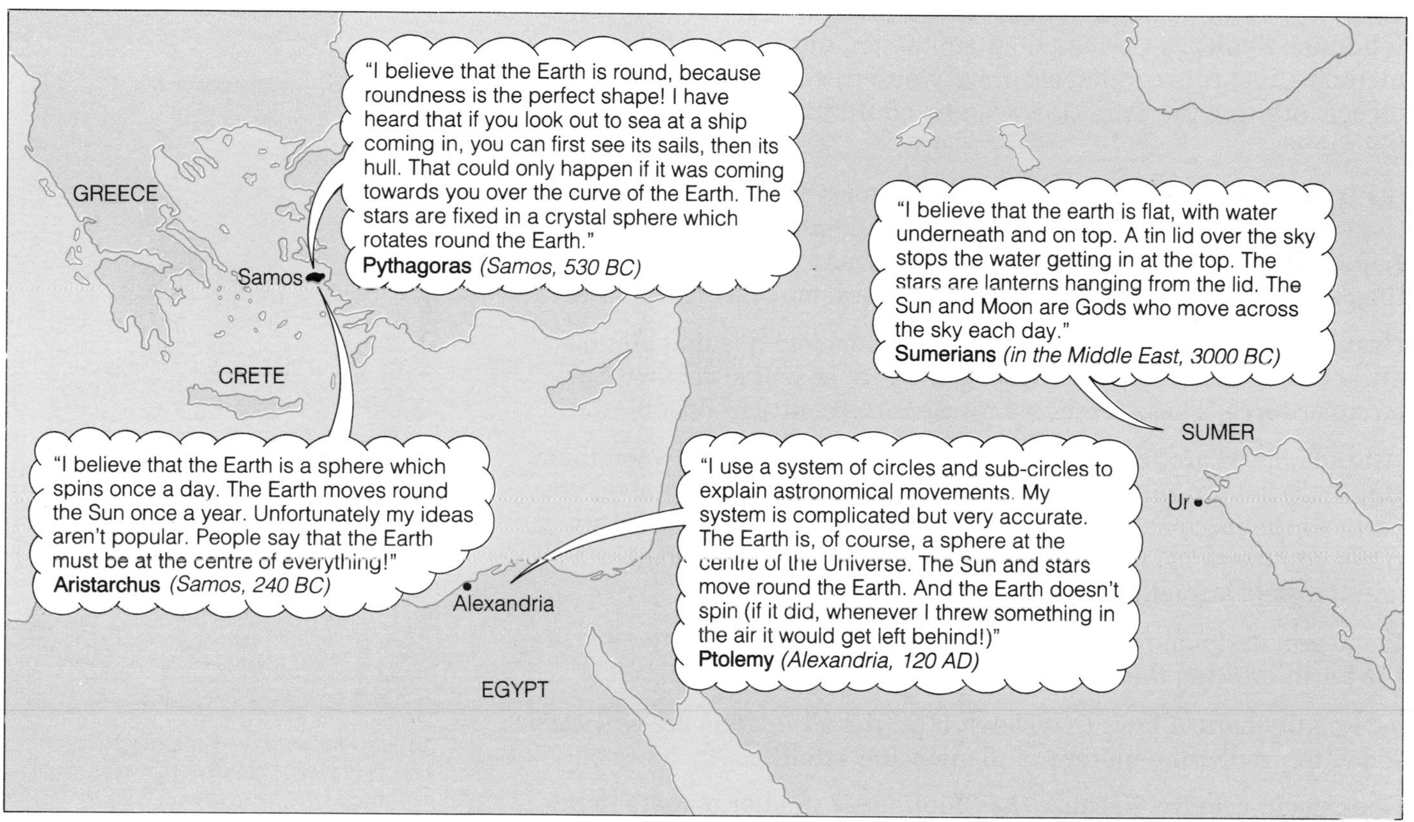

One clue to the shape of the Earth is the way the Moon looks during an **eclipse**. A lunar eclipse occurs when the Earth lies between the Sun and the Moon. The Earth's shadow falls on the Moon. This shadow is always circular.

1 Many of these speakers were simply saying what they believed. Which of them provided *evidence* in support? ▲
2 Which of the ideas was correct?
3 Make a list of the evidence that could have been used four hundred years ago, to show that the Earth is not flat. ▲
4 What additional evidence is available today to show that the Earth is a sphere?
5 If the Earth was a flat disc, what shapes could its shadow be?

Did you know?

- From the fifth to the seventeenth century AD, the Christian Church in Europe taught that the planets and stars *must* move round a flat Earth. It was felt that God would have put Earth in that most important central position.

16.1 Gravity

Gravity is the name given to the force that makes us fall down rather than up. It also keeps the planets moving in a regular pattern of orbits. Sir Isaac Newton was the first to provide an explanation for gravitational forces, but others before him laid foundations for his work.

In 1510, a Polish monk called **Nicolas Copernicus** suggested that to have the *Sun* in the centre made for a much simpler planetary system. One hundred years later, **Galileo Galilei** used the newly invented telescope to provide evidence that Copernicus' idea was correct. In 1610, Galileo saw, with his telescope, four moons orbiting round the planet Jupiter. By showing that there *were* objects which didn't go round the Earth, he showed that the Earth didn't have to be in the most important position at the centre of everything. At about this time, **Johannes Kepler**, a German mathematician, put forward three mathematical rules which accurately described the motion of all objects orbiting the Sun. However, he could find no explanation for the rules.

Sir Isaac Newton (1642–1727) collected together the thoughts of Copernicus, the results of Galileo's experiments and the work of Kepler. Using this information he put forward laws which summed up the experimental facts concerned with mass, movement and force.

Gravitation The main idea that Newton developed is that all objects attract each other. Objects with a greater **mass** will attract with a stronger **force**. These forces act towards the centre of the object.

All the objects around you have tiny attractive forces between them. But these are too small to feel. Because the Earth is so massive, you *can* feel the force between you and the Earth (when you jump up, it pulls you down again). Your own weight is the force of the Earth's gravitational attraction pulling down on you.

Large gravitational forces keep the Moon orbiting the Earth, and keep the Earth orbiting the Sun.

The mathematical laws that Newton produced in 1665 are still used today to predict the motion of planets and satellites.

Changes in gravity Because the Moon has a smaller mass than the Earth, objects on the Moon weigh less than they do on Earth. As well as depending on the mass of objects, gravity depends on the distance between the objects. The farther away from the Earth you get, the smaller is its gravitational attraction on you. Far out in deep space, you would be weightless.

1 How much do you weigh?
2 Assuming that the Earth's gravitational attraction is 10 N/kg, what is your mass?
3 About how much would you weigh on the Moon? ▲
4 What does gravitational force depend on? ▲
5 **Try to find out** a situation on Earth where you could be weightless.

Remember:

Mass is the amount of stuff in an object. It is measured in kilogrammes (kg).

A **force** can cause a mass to accelerate. Forces are measured in newtons (N).

Weight is the force due to gravitational attraction. Because weight is a force, it is measured in newtons.

An astronaut trying to drink orange juice without gravity. What other problems might this bring?

Did you know?

- A bag of sugar has a mass of 1 kg. On Earth it weighs about 10 newtons.
- An astronaut with a mass of 60 kg weighs about 600 N on Earth, but only 100 N on the Moon. Out in space, far from any other objects, an astronaut would be weightless.

16.2 Days and years

Starting off

The Earth's spin

The Earth spins on its **axis**, once in about 24 hours (1 day).

The side of the Earth nearest the Sun is lit up by the Sun's light. The other side of the Earth is in darkness.

As the Earth spins, the figure A moves from the dark part (night time) into the light part (day time).

Our days and nights are caused by our part of the Earth spinning into, and out of, the Sun's light.

The Earth's movement round the Sun

The Earth also moves round the Sun. It **orbits** the Sun once every year.

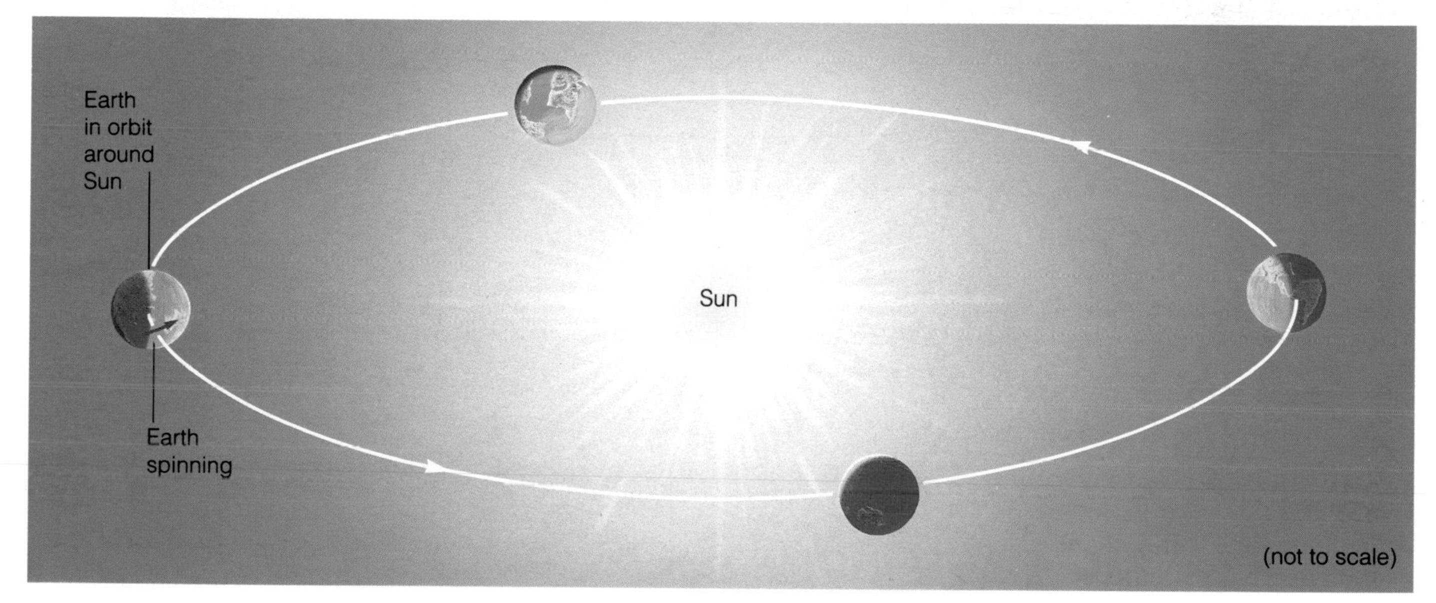

Our standard year has 365 days. But it takes the Earth 365¼ days to make a complete orbit round the Sun. At the end of four years of 365 days, the Earth would still have another day's orbiting to do, to get back to its original position. So every fourth year an extra day (February 29th) is added on to the calendar. This is called a **leap year**.

1. Explain how day and night are caused by the Earth's spin. A drawing might help. ▲
2. How long does the Earth take to orbit the Sun? ▲
3. 1988 and 1992 were leap years. Which years are the three leap years after 1992?

Did you know?

- Actually, the time for one complete orbit is 365 days, 5 hours and about 48½ minutes. That's slightly *less* than 365¼ days. So a further slight adjustment is necessary! Leap years are missed out in century years (1700, 1800, etc.), *unless* the date can be divided exactly by 400. So 2000 *is* a leap year.

16.2 Day and night

Going further

Different lengths of day and night

The **axis** on which the Earth spins is an imaginary line through the north and south pole.

This axis is at a **tilt** to the rays from the Sun to the Earth.

In the picture, three lines have been drawn around the Earth. These lines are called **latitudes**. The latitude at the middle of the Earth is called the **Equator**. At the Equator, there are always 12 hours of darkness and 12 hours of light. Because of the Earth's tilt, other latitudes have a daylength which changes through the year

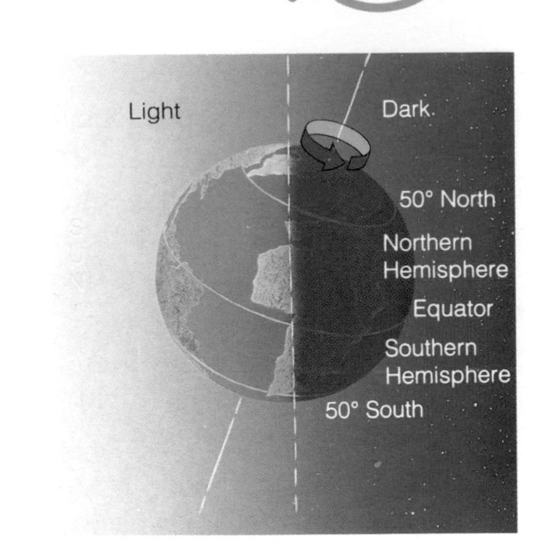

The position of the Earth in June and December is shown here. The Earth's tilt is always at the same angle. Drawing the path taken during one day's spin, shows two things:

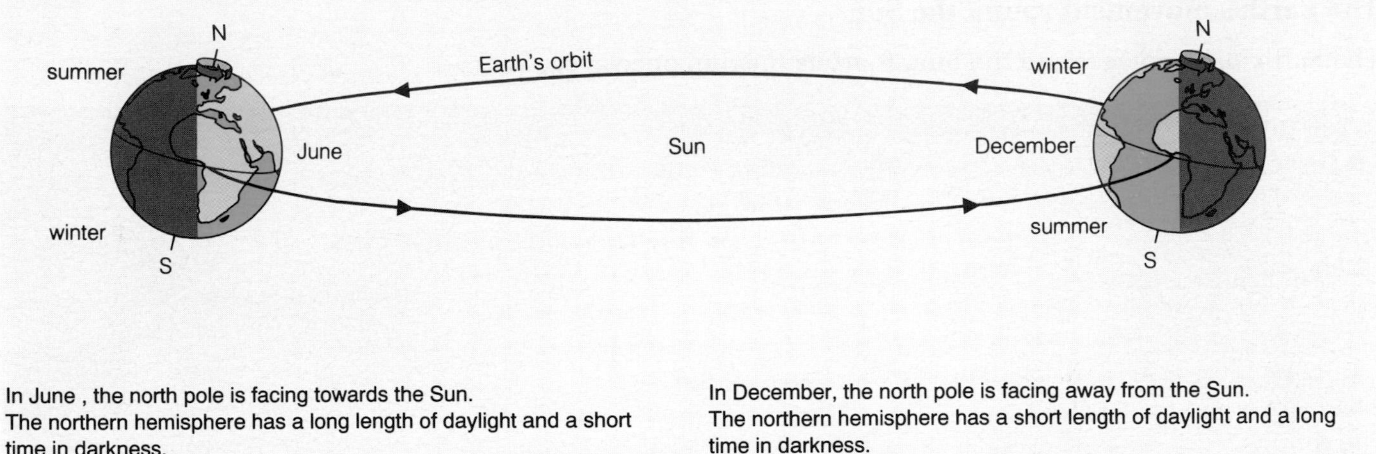

1. Use the drawing to explain why the northern hemisphere has long nights in December and short nights in June. ▲
2. How will the lengths of day and night vary in the southern hemisphere?
3. In March and September the Earth's axis does not tilt towards or away from to the Sun. What does this tell you about the length of day and night at these times of the year?
4. **Try to find out** where Spitsbergen is and explain why the Sun shines at midnight there in June.

Spitsbergen at midnight in June. At this time of the year the Sun never disappears.

Did you know?

- The Earth's axis points to a position in the sky near to the pole star (Polaris). However, the Earth 'wobbles' slightly on its axis, rather like a spinning top which is slowing down. It takes 25 000 years for one wobble. In 12 500 years the pole star will not be over the Earth's north pole, but it will be back in place in 25 000 years!

16.2 Hot and cold

For the enthusiast

Seasonal changes

In New Zealand, Christmas is in mid-summer, and June's the time for skiing. This is another result of the tilt of the Earth's axis.

Imagine two identical beams of sunlight falling on your part of the Earth, one in June and one in December.

Each beam will contain the same amount of heat energy.

Christmas shopping in Sydney

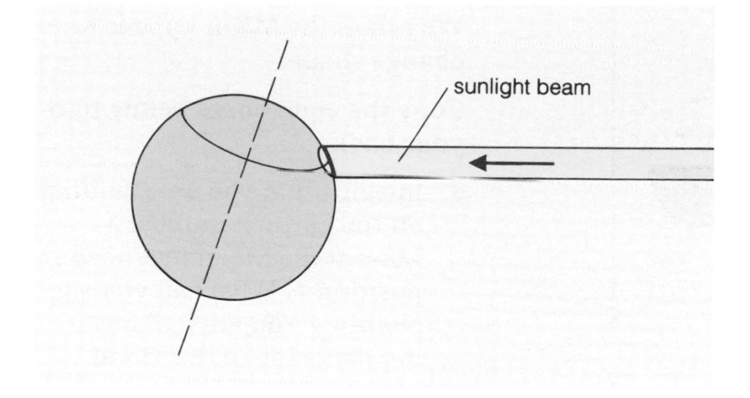

In the northern hemisphere in June, you can see that the Sun will be higher in the sky.

Your beam of sunlight is concentrated onto a small area of the Earth's surface (you can see this in the diagram).

The ground heats up because of this concentration of the Sun's heat energy.

In June, the northern hemisphere gets warm.

In December in the northern hemisphere, you can see that the Sun will be lower in the sky.

Your beam of light is now spread out over a much larger area of the Earth's surface (you can see this in the diagram).

Because the Sun's heat energy is spread over a larger area, the ground doesn't heat up so much.

In December, the northern hemisphere is cool.

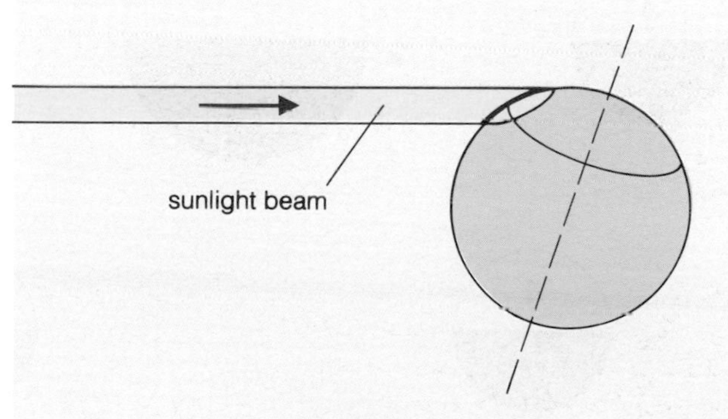

1. Explain how the Earth's tilt produces the variations in winter and summer temperatures. ▲
2. Explain why summer comes in June in the northern hemisphere and in December in the southern hemisphere.
3. Why is the area close to the equator hot all the year round?
4. Design a model to show how the heating effect of the Sun changes when the beam falls at different angles. You could use a ray box and a shield with a 1 cm square hole to produce the 'sunbeam'. How could you detect the heating effect at different angles?
5. **Try to find out** what happens at the north and south poles during winter and summer.

Did you know?

- In its orbit, the Earth gets closer to, and farther away from the Sun.
  This doesn't have much effect on the climate, though. In June, the Sun is about 152 million km away, and in December, about 147 million km.

16.3 The Moon's phases

Starting off

The Moon's shape seems to change from 'full' to 'half' in about a week. A week later it can't be seen at all. But a week after that, it appears to grow again.

The Moon moves around the Earth once in roughly 28 days. We call this period a **lunar month**. The Sun lights up the side of the Moon which is facing it. We see the side of the Moon which is lit up. The different shapes of the Moon which we see are called the **phases** of the Moon.

The drawing shows how Earth and Moon are lit up by the Sun. The Moon is shown in 8 positions during the lunar month. It takes about 3½ days for the Moon to travel from one position to the next.

Did you know?

- When the Moon shows as a tiny crescent, you can sometimes just see the dark side of the Moon. When this happens, the Sun's light is being reflected back from the Earth's surface to fall on the Moon.

Why does the Moon appear to change shape?

Copy the eight boxes below into your book.

1. Imagine that you are standing on the Earth at number 1. Look at the Moon overhead in position 1. What can you see? Nothing! You are looking at the dark side of the Moon.

   Draw this in your number 1 box, as shown.

2. Next, turn the book so that Moon 2 is at the top. Imagine that you are standing on the Earth at number 2. Look at the Moon in position 2. What can you see? Just the right side is lit up! (But remember that you're trying to represent a spherical Moon.) Draw this in your number 2 box, as shown.

3. Now do the same for the other positions.

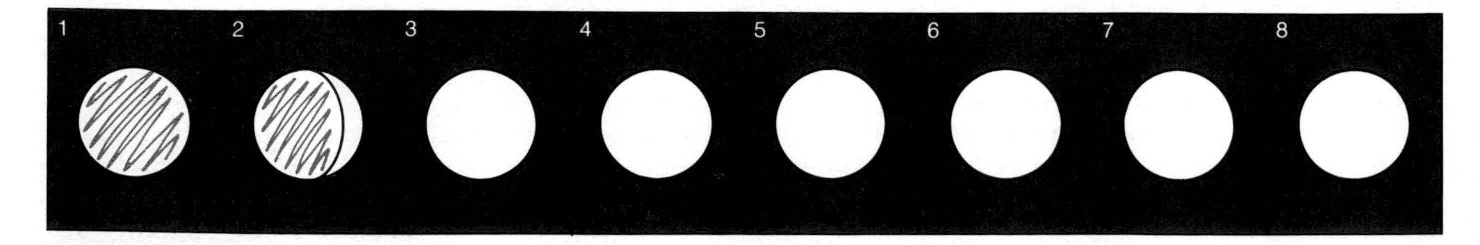

16.3 Shadows in space

Going further

The Sun lights up the Earth and the Moon, and casts shadows in the space at the far side of them. When the Moon passes into the Earth's shadow a **lunar** eclipse occurs. The Earth blocks out the Sun's light. A **solar** eclipse occurs when the Moon's shadow falls on the Earth. The Moon blocks out the Sun's light.

Warning You must never look directly at the Sun.

Eclipses of the Sun and Moon

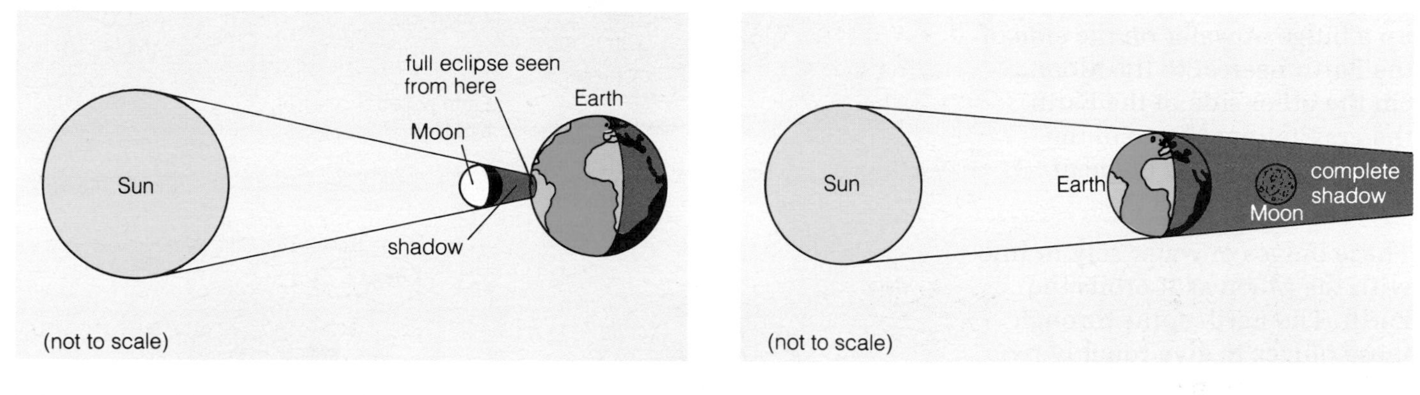

Why are eclipses so rare?

Since the Moon orbits the Earth once in roughly 28 days you might think that eclipses should happen every two weeks (at new moon and full moon). But although the Earth, Moon, and Sun are *roughly* in line every fortnight, they are rarely *exactly* in line. There are two reasons for this.

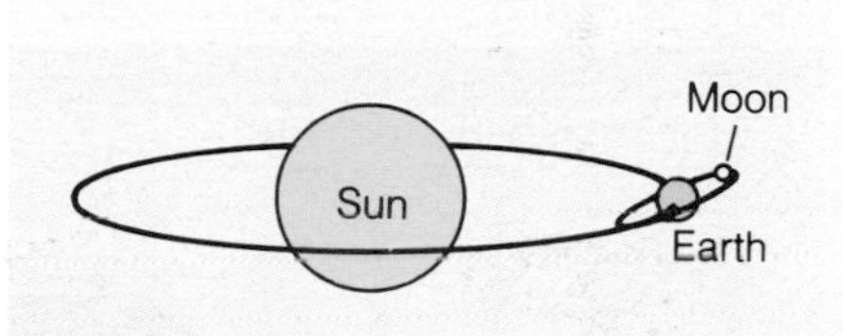

(not to scale)

The first is that the Moon's orbit is at an angle to the Earth's orbit.

For the second, you need to look at the sizes of the Sun, Moon and Earth, and at the distances between them. The drawing below shows, to scale, the size of the Earth and Moon, and their distance apart. On the same scale, the Sun would be a 50 cm ball, about 50 metres away. The distances are so large, and the objects so tiny, that the alignment has to be very precise for an eclipse to occur.

Earth

Moon

to scale

Answer questions **1** to **4**,
a) for a solar eclipse, b) for a lunar eclipse.

1 What is being eclipsed? ▲

2 Is the eclipse at night or day? ▲

3 Is the moon 'new' or 'full'? ▲

4 Is the eclipse seen from just a few places, or from half the Earth?

5 **Try to find out** when the next total eclipse of the Sun can be seen from this country.

Did you know?

- **Sizes and distances** (approximate)

Diameter of Sun	1 400 000 km
Diameter of Earth	12 750 km
Diameter of Moon	3 500 km
From Sun to Earth	150 000 000 km
From Earth to Moon	400 000 km

16.3 Water on the move

For the enthusiast

Tides

Tides are caused by the **gravitational force** of the Moon (and to a smaller extent, of the Sun) acting on the Earth's seas.

Gravitational forces are larger for things that are nearer to each other. So the Moon's force pulls up a bulge of water on the side of the Earth nearest to the Moon. On the other side of the Earth, the gravitational force of the Moon is less, and a bulge of water is 'left behind'.

These bulges of water stay in line with the Moon as it orbits the Earth. The Earth spins through these bulges to give roughly two high tides each day.

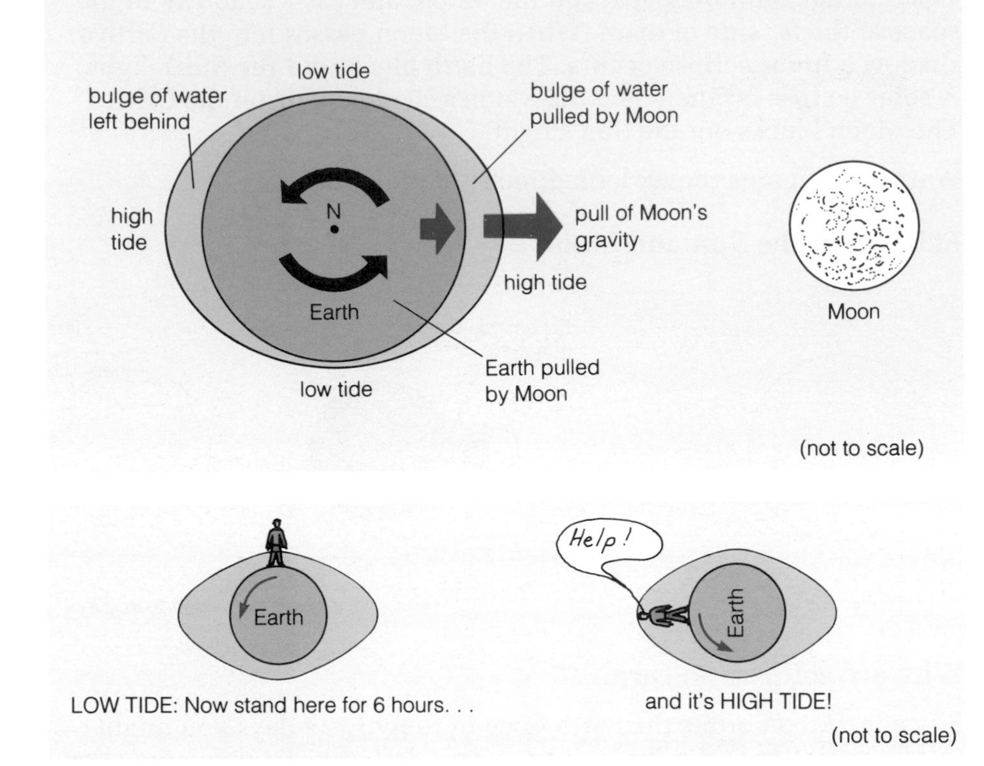

Spring tides and Neap tides

When the Moon and Sun are in line, as at new moon and full moon, their gravitational forces add up to give very high and very low tides. These are called **spring tides** (in the sense of '*spring*ing' up!).

When Moon, Sun and Earth are at right angles, there is a smaller range between high tide and low tide levels. These are called **neap tides**.

1 What causes the bulges of water on the Earth's surface? ▲
2 Why are there roughly two high tides a day? ▲
3 Why do spring tides occur near the time of a full or new moon? ▲
4 Explain how the Moon's movement round the Earth will make each tide later day by day (use the large drawing as a starting point).

Did you know?

- At high spring tides, the River Severn has a **tidal bore**. The shape of the estuary funnels the incoming water into a wave 2 m high which travels up river at 20 km/hr.

16.4 Impossible journey 1

Starting off

A journey through space

Imagine that you are on **Earth**, ready to set off towards the outer regions of the Universe. Remember, though, that this is an impossible journey! Even travelling at the fastest possible speeds it would take you more than 10 000 million years.

On the first stage of the journey you leave Earth and travel just beyond the **Solar System**.

Leaving Earth

As you leave the Earth's atmosphere, the sky darkens to black. The Earth now appears as a bluish sphere. In its atmosphere there are swirling white clouds of water vapour.

A large **satellite**, more than a quarter of the Earth's diameter, is in orbit round the Earth. This is the **Moon**. Its dry, dusty surface is covered with craters. The large grey regions you can see on the Moon's surface were once thought to be seas. They are now known to be low-lying areas which became covered with molten lava.

You can see that both the Earth and Moon are lit up by the Sun.

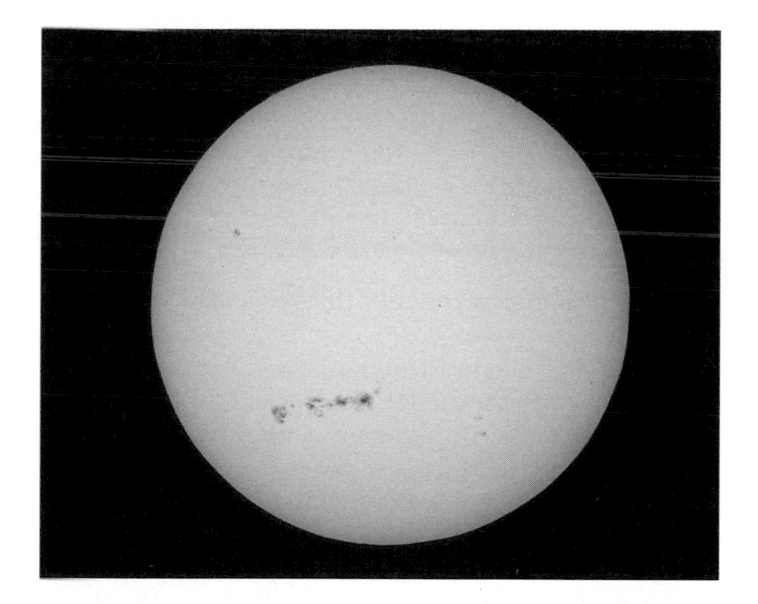

The Solar System

Our **Sun** is a star. It is a hot ball of gas, fuelled by nuclear reactions. It gives out heat and light. You must never look directly at the Sun as it is too bright.

Around the Sun you can see a number of smaller, cooler objects in orbit, all lit up by the Sun. Their orbits are all more or less in the same plane. There are nine of these **planets**, and many thousands of smaller particles called **asteroids**. Our Earth is the third planet out from the Sun.

The surface temperature of the Sun is normally about 6000°C. From time to time, patches on the surface cool down to about 4000°C. These cool patches appear darker. They are called **sunspots**. Can you see them in the photograph?

1 What is the Earth's nearest neighbour in space? ▲
2 How many planets orbit the Sun? ▲
3 How many planets are closer to the Sun than Earth? ▲
4 Which is the brightest object in the solar system? ▲
5 **Try to find out** what effects sunspots can have on Earth.

Did you know?

- At spaceship speeds it takes about 3 days to get to the Moon. At the same speed it would take about a year to travel the distance to the Sun.

16.4 The Solar System 1

Going further 1

Saturn. The rings are made from millions of fragments which are about 10cm across.

The first four planets, Mercury, Venus, Earth, and Mars, are small and rocky. The next four planets, Jupiter, Saturn, Uranus, and Neptune are larger, and gaseous. Pluto, the outermost planet, is small and rocky.

The two planets nearest to the Sun take less than a year to make one orbit. Planets with larger orbits take longer to make one complete orbit.

Did you know?

- The planets' orbits round the Sun are not quite circular. They are ellipses (slightly 'squashed' circles). Normally Pluto is the farthest planet from the Sun. However, its orbit is quite noticeably 'squashed'. Between 1979 and 1999 Pluto is actually closer to the Sun than Neptune is! (The scale distance now would be about 1475 m.)

A Scale Model

If you tried to make a scale model of the Solar System, and used a large beach ball to represent the Sun, then Pluto would be about 2 km away.

Here is a table of scaled-down sizes and distances.

	Size (diameter)	Distance from Sun
SUN	47 cm	–
Mercury	2 mm	20 m
Venus	4 mm	35 m
Earth	4 mm	50 m
Mars	2 mm	75 m
Jupiter	47 mm	250 m
Saturn	40 mm	500 m
Uranus	17 mm	1000 m
Neptune	17 mm	1500 m
Pluto	2 mm	2000 m

	Size	Distance from Earth
Moon	1 mm	13 cm

1. Which planet is roughly the same size as Earth? ▲
2. Which is the largest planet? ▲
3. On the scale on the left, how much of the model could be fitted into a) your school hall, b) your school playing field?
4. What would be the disadvantage of reducing the scale by 10, so that the whole model solar system would fit in a smaller space?
5. Most of the planets we can see from Earth are named after ancient deities. Which is the odd one out?

Try this Make models of Sun, Earth, and Moon. Use them to show how the Earth spins as it orbits the Sun, with the Moon orbiting the Earth at the same time. You will have to decide on a scale for times.

16.4 The Solar System 2

Going further 2

	Mercury	Venus	Earth	Mars	Jupiter	Saturn	Uranus	Neptune	Pluto
Average distance to Sun (million km)	60	108	150	230	780	1400	2900	4500	5900
Revolution period ('year') (Time for one orbit) (in years or days)	88 d (=0.25 y)	225 d (=0.6 y)	365.25 d (=1 y)	687 d (=1.9 y)	12 y	29 y	84 y	165 y	248 y
Rotation period ('day') (Time for one spin) (in days, hours, minutes)	59 d	243 d	23h 56min	24.5 h	10 h	10.25 h	11 h	16 h	6.5 d
Diameter (km)	5 000	12 000	12 750	7 000	140 000	120 000	52 000	50 000	5 800
Surface	cratered rocky	rocky + clouds	rocky + water + clouds	cratered rocky	gaseous outer layers	gaseous outer layers	gaseous outer layers	gaseous outer layers	rocky
Average temperature (°C)	350 day –170 night	480 at surface	22	–23	–150 at cloudtop	–180 at cloudtop	–210 at cloudtop	-220 at cloudtop	-230
Atmosphere (main gases)	none	carbon dioxide	nitrogen oxygen	carbon dioxide argon	hydrogen helium	hydrogen helium	hydrogen helium methane	hydrogen helium methane	?
Number of known satellites ('moons')	0	0	1 (the Moon)	2	>20	>20	5	8	1

1 What is the Earth's satellite called? ▲

2 Which planet has a day which is a similar length to Earth's? ▲

3 How long are Earth's 'day' and 'year'? ▲

4 What is unusual about the lengths of Venus's 'day' and 'year'? ▲

5 Why do two temperatures need to be shown for Mercury?

6 What has Earth's atmosphere got that no other planet has, that enables us to live? ▲

7 If you could take an unlimited supply of oxygen, which planet would offer the best chance of survival? Explain what factors you took into account in making your decision.

8 For the first 5 planets, draw a line graph to show how **revolution period** varies with ***distance* from the Sun**. Draw the *period* on the y-axis, (vertical). Use a scale of 2 cm for each year. Draw the *distance* on the x-axis (horizontal). Use a scale of 2 cm for 100 million km. Draw a line through your points to see the pattern.

9 A group of small bodies called asteroids goes round the Sun at an average distance of about 400 million km. Use your graph to estimate how long they take to make one orbit.

10 The two **Voyager** spacecraft have made a number of new discoveries about the planets. **Try to find out** about some of these.

16.4 A new planet discovered

For the enthusiast

You are the captain of a space craft and report a new discovery to Mission Control:

"We have collected information on a newly discovered planet. For their own safety, all Earth people must read this information before attempting to land".

British astronaut Helen Sharman.

GALAXY STAR FLEET NEW PLANET REPORT

Distance from central star	180 million kilometres	
Period of revolution	500 days	
Rotation period	48 hours	
Diameter	25 000 kilometres	
Surface	Rocky, with some high mountains	
	Water	
	Clouds	
Mean temperature	12 °C	
Atmosphere	60 % nitrogen	30 % oxygen
	8 % carbon dioxide	2 % other gases
Mass	Estimated at about 8 times the Earth's mass	
Magnetic field	None	
Further notes	1 Some small coal deposits have been noted	
	2 There are small polar ice caps	
	3 Chlorophyll molecules have been detected	

Pre-landing check You may need to refer to the previous page to make comparisons with conditions on Earth.

1 The atmosphere is rich in oxygen.
   a) Will your crew be able to breathe this atmosphere?
   b) How will the change affect their bodies?
2 Explain why there might be a risk of fire on landing.
3 Look at the rotation period. Will you need to plan special sleeping arrangements for your crew?
4 The larger mass and size of this planet mean that gravitational forces will be about twice those on Earth.
   a) What effect has this on the weight of your crew?
   b) How will it affect the amount of fuel you will need to take off?
5 The report mentions two things necessary for life as we know it.
   a) What are they?
   b) What other information is given to show that life exists on this planet?
   c) How long do you think plants might have existed on this planet? (You could use your knowledge of the formation of coal on Earth to help estimate a minimum time.)

Did you know?

- The Greeks, 2000 years ago, knew the planets Mercury, Venus, Mars, Jupiter and Saturn.
- Uranus was discovered in 1781, and Neptune in 1846.
- Pluto's existence was predicted in 1905, but it wasn't discovered until 1930. Its image was spotted in different positions in two separate photographs.

16.5 Impossible journey 2

Starting off

Your journey continued

In the next stage of your impossible journey you travel from the Solar System to the outer edge of the Milky Way galaxy.

Stars

As you leave the Solar system, the planets become invisible, but you can still see the Sun. The Sun is one of many stars grouped together in one of the spiral arms of our **galaxy**.

On a dark night you may already have seen the stars in this spiral arm as a misty area across the sky, known as the **Milky Way**.

But you will have to travel much further before you can see what the whole galaxy looks like.

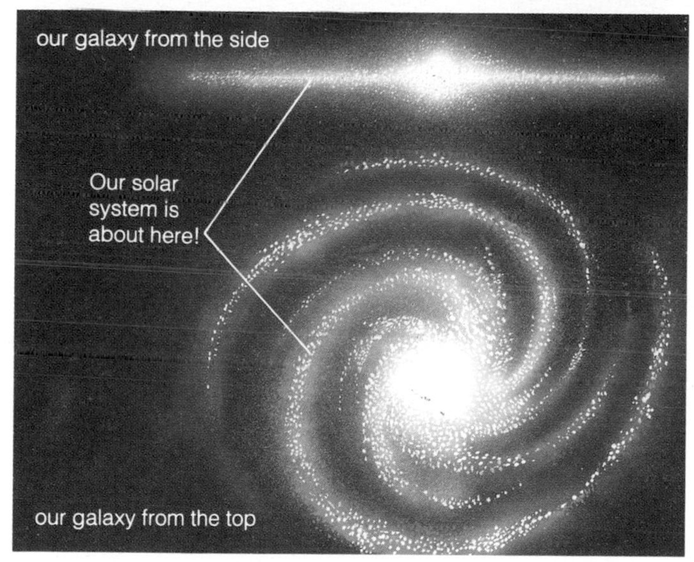

The 'Milky Way' galaxy

Our galaxy contains about 100 000 million stars. It is spiral in shape. It has a central core of older stars, and the spiral arms contain newer stars.

These are drawings, *not* photographs, of the Milky Way galaxy. (Nobody has every been far enough away to take a picture of our own galaxy from the outside!). One 'view' appears 'edge on', and the other is a 'top view'.

The Sun would be about two thirds of the way out from the centre of the galaxy.

1 Describe the shape of our galaxy. ▲
2 Roughly how many stars are there in our galaxy? ▲
3 Make a drawing of a spiral galaxy and mark a suitable position for the Sun. ▲
4 Why is it impossible to take a picture from the outside of our own galaxy? ▲
5 As you leave the solar system, why can you still see the Sun, but not the planets?
6 **Try to find out** the name of the nearest star to the Sun. (It is about 4 light years away.)

Did you know?

- The stars in a galaxy swirl round the centre. The Sun takes 225 million years to make one orbit round the centre of the galaxy.
- Our galaxy is about 1 million million million km across!
- Astronomers have defined a unit of distance (called the light year) to cope with these vast numbers. A **light year** is the distance light travels in 1 year. One light year is about 10 million million km.

16.5 Impossible journey 3

Going further

Your journey continued

In the final stage of your impossible journey you travel from the Milky Way galaxy, through the local group of galaxies and on to the outer edge of the Universe.

The 'Andromeda' group of galaxies

Our **galaxy** is one of a group of galaxies. There are about 20 galaxies in this **local group**. The group is named after the largest member of the group, the **Andromeda** galaxy.

The Andromeda galaxy is about 130 000 light years across.

(If you know just where to look, you can see the Andromeda galaxy from Earth, without a telescope. It takes up about the same space as the Moon in the sky, but it is very, very faint.)

. . . and beyond

The **Universe** contains lots of empty space, lots of gas molecules and thousands of millions of clusters of galaxies.

The farther out you go, the more spaced out things are!

1. How many galaxies are there in our local group? ▲
2. How many light years is the diameter of the Andromeda galaxy? ▲
3. About how many kilometres is this?
4. Write your 'universal address'. Start with your normal address, and add information which locates the Earth's position in the Universe. You will need information from the whole of your impossible journey.
5. Explain why a journey to the outer regions of the Universe is impossible.

Did you know?

- The distance to the Andromeda galaxy is 2¼ million light years. That's more than 20 million million million km!
- The distance to the Sun is only 150 000 000 km. It takes light 8 minutes to travel this distance.

16.5 Once upon a time

For the enthusiast

How the Universe developed

Our Universe began more than 10 000 million years ago. Most scientists think that, in the beginning, all matter was packed into one dense mass. There was a mighty explosion! This '**big bang**' shot out all the matter that now exists. The Universe has continued to expand ever since.

The Solar System forms

In our part of the Universe, the Solar System formed about 4 600 million years ago. A large, swirling mass of gas and dust started to collect due to the force of gravity. Most of the particles joined together to form the **Sun**. Strong gravitational forces packed the particles tightly together. Nuclear reactions started. The Sun began to give out heat and light energy.

The rest of the particles joined together to form smaller, cooler objects, called **planets**.

Progress on Earth

When **Earth** first formed, enough heat was produced to melt the surface. It took about about 500 million years for the hot, liquid rocks to cool down and become solid. Over the next 500 million years, the Earth's surface became covered with water. By 2000 million years ago, algae-like plants had developed. These used the Sun's light to produce oxygen.

Now that the Earth had **water** and **oxygen**, other forms of life could start to develop.

1 About how long ago did the Earth become solid? ▲
2 What kind of plants were the first to start producing oxygen? ▲
3 What two things did animals need before they could survive on Earth? ▲
4 You have volunteered to be the 'science correspondent' for your local primary school newspaper. Write a short report (about 200 words maximum) describing how the Earth formed and developed. ▲

Universe forms more than 10 000 million years ago

10 000 million years ago

5 000 million years ago

Solar system and Earth form

Oldest known rocks

Primitive cells develop

Algae-like plants

Single celled animals

Colonies of cells (?sponges)
Shellfish, then fish, ferns
Amphibians, insects, reptiles
Mammals, then birds, flowers
NOW

Did you know?

- Light travels at 300 000 km/s.
  The Sun is 8 '**light minutes**' away. When you see the Sun, the light that enters your eye set off 8 minutes before.
- The star Sirius is the nearest star we can see from the northern hemisphere (apart from our Sun). Sirius is 9 **light years** away. The light you see from Sirius set off 9 years ago.
- When you look at the stars, you look into the past!

16.6 "A vast leap for mankind"

Starting off

Man on the Moon

On 16 July 1969, a **Saturn V** rocket carrying the **Apollo 11** spacecraft lifted off from Cape Canaveral in Florida. Three astronauts, Neil Armstrong, Edwin Aldrin and Michael Collins were on board. After orbiting the Earth twice, they set off on their three day journey to the Moon. After a day in lunar orbit, the lunar module, with Aldrin and Armstrong inside, separated from the orbiting 'Command and Service Module' to land on the Moon.

Because of gravitational effects between the Earth and Moon, the Moon always has more or less the same side towards the Earth. Each lunar month, each point on the Moon's surface has a fortnight's darkness and a fortnight's light. Temperatures range from -150°C in darkness to 120°C in daylight. The astronauts' spacesuits have to provide for this temperature variation.

Voyaging onwards...

In January 1990, the unmanned **Voyager 2** spacecraft reached the outer limits of the Solar System. As it approached Neptune, gravitational forces increased its speed to 100 000 km/hour.

As it passed the planet it transmitted new and exciting information back to Earth. But how long had that journey taken? *Over 12 years!*

...and onwards

Voyager is now carrying on, out of the Solar System and into space. It carries a plate on the outside which attempts to give information about us and Earth, to anyone who may read it! There is a problem, though, with this method of communication! It will take about 100 000 years for Voyager to reach even the nearest star beyond the Solar System.

Neptune

1. Two of the Apollo 11 astronauts landed on the Moon. What was the third doing at that time? ▲
2. For what reasons was the Voyager 2 journey not suitable for astronauts to go on?
3. **Try to find out** more about manned space flights to the Moon.

Did you know ?

- Craters are found over much of the Moon's surface. They were caused by meteorites crashing into the Moon just after its formation many millions of years ago.

16.6 Space travel limitations

In this chapter you have taken part in a three-stage 'Impossible Journey', and you have 'visited' a newly discovered planet orbiting another star. This is, unfortunately, **science fiction**!

Your scale model of the solar system spread about two kilometres from the model Sun. If you added stars to this, the *nearest* star to the Sun would be around 15 000 km away. Compared to that 15 000 km, humans have only actually managed to travel the 13 cm to the Moon.

Our space-travelling sci-fi heroes and heroines manage to cross the Universe in minutes. But even on your scaled down model, the Universe would be perhaps 100 million million kilometres across. Clearly, *real* humans have a long way to go before they can start to imitate the 'space travellers'!

But it isn't just that we have a lot of development to do before science *fiction* is turned into science *fact*. There are very definite limitations as to what is possible.

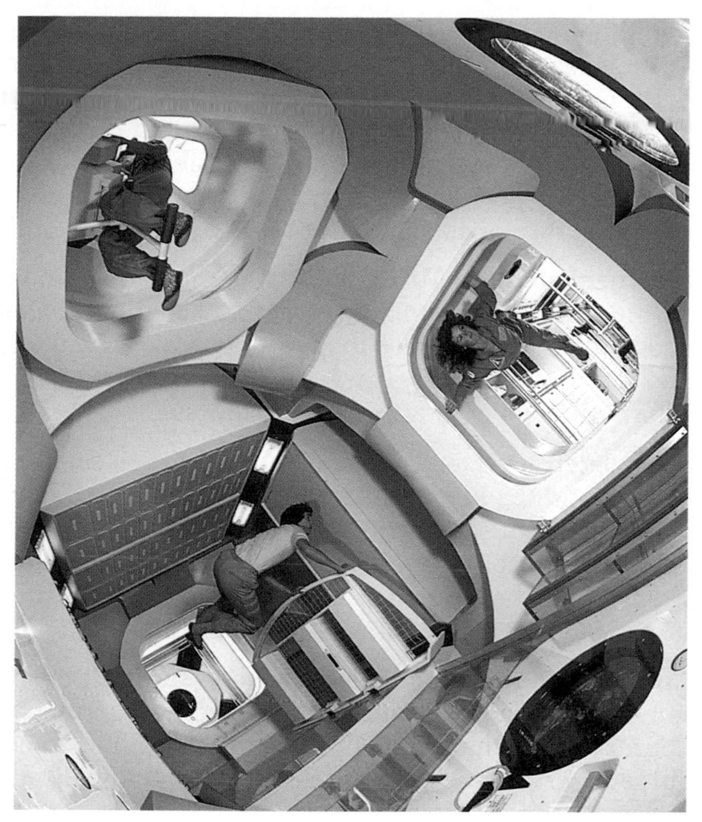

Inside a space station.

What *is* possible?

There are limits to whatever humans do. For example, they can't lift ten tonnes of coal or run at 100 km/hour.

In 'real' space travel, there are also limitations. These include:

Speed It is thought that it is impossible to travel faster than the speed of light (that's about 300 000 km/s). Even at this speed it would take you four years to get to the nearest star beyond the solar system. At Voyager's *maximum* speed of 100 000 km/hour (about 30 km/s) it would take 40 000 years to get to there! And it would take a million times longer to get even to some of the 'nearby' galaxies!

Time You are not likely to live for more than 100 years. (And remember that 'civilised' human beings have been around for only 10 000 years - that's about 400 generations!)

Capacity There would be limits to the number of people and the amount of supplies you could take with you on a journey.

Did you know ?

- Radio messages travel at the speed of light.

1 What are the major obstacles to 'real-life' long distance space travel? ▲
2 Suggest how some of these problems could be solved (but try to be realistic!).
3 You could extend the length of a journey by planning it to last over several generations. What problems might this cause?
4 Why would communicating with Earth become more of a problem the further you went?
5 Decide on what you think is the furthest place that humans could reasonably reach. Describe the plans that you would need to make for this voyage.
6 Think of some of the travels which popular sci-fi characters make. Try to explain why their exploits are, in fact, impossible.

16.6 Satellites

Going further 2

The Moon is the Earth's natural satellite, but there are plenty of artificial satellites too. Like the Moon they are held in orbit around the Earth by gravitational force. Because their orbits are smaller, satellites close to the Earth take less time to go round it than those further away. At a height of 160 km, one orbit takes about one hour and a half. Some satellites are just discarded rockets from space launches. Others are put into orbit for special purposes.

Weather satellites 'Tiros', the first of many civilian weather satellites, was launched in 1960. Some weather satellites are positioned far above the Earth's surface to give an overall view of cloud formations and movements. Others orbit closer to the Earth to give more detailed information on clouds and temperatures.

Land satellites 'Landsat' is one of many such satellites. It is in close orbit round the Earth and has an orbit time of 1¾ hours. As it passes overhead, Landsat views a strip of the Earth's surface. The information collected is used to study such things as water pollution, the effect of industry on the environment, and to distinguish different types of crops. Landsat images are also used for making maps.

Communications satellites are used to transmit telephone calls, television broadcasts and other electronic information from one country to another. Global positioning satellites transmit data so that ships and aircraft can locate their position to within 100 metres. Both are in geostationary orbits.

Two types of satellite orbit are particularly useful.

Polar orbits

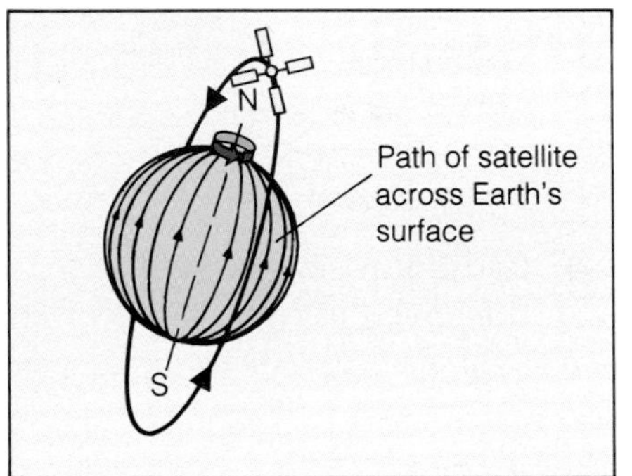

Polar orbits
The satellite is in orbit over the Earth's poles. It travels from north to south around one side of the Earth, and then back north round the other side. However, while this is happening, the Earth is spinning beneath the satellite. The result is that the satellite appears to travel a spiral track over the Earth's surface.

Geostationary orbit

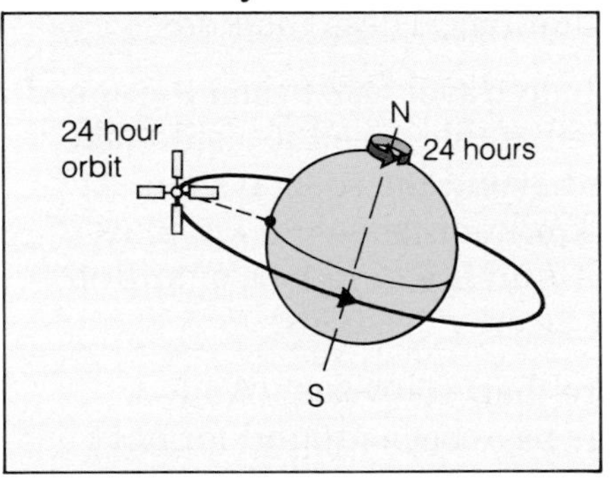

Geostationary orbit
The satellite travels eastwards in an orbit directly above the equator. Launched to a height of 36 000 km, it takes one day to make one complete orbit. However, the Earth is spinning beneath the satellite in the same direction. The result is that each satellite appears to remain stationary above a point on the equator.

1 Make a list of the things artificial satellites are used for. ▲
2 How does the time of one orbit vary with the height of a satellite? ▲
3 Explain how a satellite can be stationary above the Earth's surface. ▲
4 **Try to find out** about the various satellites which are used for space research. (The 'Hubble' telescope is just one example.)

Did you know?

- You can see artificial satellites as star-like points of light travelling quite slowly across the background of the stars. They are lit up by the Sun's light. The best time to look for satellites is just after dark in the evening, or just before sunrise, though in summer you can see them all night. During one half hour you might see three or four satellites. A satellite often travels across the sky for a few minutes before slowly fading in brightness as it passes into the Earth's shadow.
- The French 'Spot' satellite takes a picture of your street once every 26 days.

16.6 Astronomy in disorder

Finishing off

You are preparing a poster. It is about the way people have thought about astronomy, and about their uses of astronomy.

But tragedy has struck! Your poster got moved before you had chance to paste each piece of information on it. Now your paragraphs are in a muddle.

Read the paragraphs carefully. Decide which is the correct sequence. Then write down the paragraph letters, in the correct order. The calendar will help.

A The Greeks were much more sophisticated. They put a round Earth at the centre of the Universe. The Sun, Moon, planets and stars all went round the Earth. They imagined all sorts of complicated mechanisms to try to explain the planetary movements they saw.

B About the same time, Stonehenge was being constructed in England. It could be used to predict eclipses.

C Our ape-like ancestors were developing on Earth more than a million years ago. By 200 000 years ago they were recognisably human. But even 20 000 years ago, stone age people were mainly wandering food collectors.

D Copernicus guessed that the Sun, and not the Earth, was at the centre of the Universe. This made the movements of the planets much simpler to explain. The Church, however, could not believe that God had arranged things so that the Earth was not in the most important position. "Everything", they said, "must go round Earth".

E 'New' stone age people marked the beginning of civilisation. By 10 000 years ago they were living in fixed villages and were planting seeds to grow crops. To do this successfully, they needed to look at the Sun, Moon and stars to plan a calendar of planting times. These were the first astronomers!

F Modern research has shown that the Sun is not at the centre of the Universe. It is merely at the centre of the Solar System.

G In 1610, Galileo saw through his new telescope that 4 'moons' were orbiting round Jupiter. "So everything *doesn't* have to orbit round the Earth!", he explained. Copernicus was proved right!

H By the 1700s, accurate astronomical measurements were being used to make ocean navigation possible and safer.

I Later, Newton's ideas on gravitation explained **why** the planets moved as they did.

J By 2000 BC, the Babylonians in the Middle East were making quite accurate astronomical measurements. They used these for calendar making and for astrology. Schools of astronomy were also set up in China to study and map planets and stars.

CALENDAR

'New' stone age. Using astronomy to plan planting times.
More than 10 000 years ago
10 000 years ago

5 000 years ago

Sumerians
Babylonians
Stonehenge (≈2000 BC)

Greeks
(BC > AD)

Copernicus, Galileo, Newton
Astronomy for navigation
NOW

Healthy living 17

Human beings in developed countries are brought up to think that the average length of their life (or **life expectancy**) will be 'three score years and ten' (that's 70 years).

But you will all know someone who has lived to be much older than that. And, sadly, many of you will have known someone who died at a much earlier age.

Even apart from accidents, there are many reasons why some people live longer than others.

Your family history seems to have a part to play. So does the type of food you eat, your lifestyle, how well your body fights disease, and whether you are male or female.

The scientific progress that our society has made in the field of nutrition, and in the prevention and cure of disease, means that we all tend to live longer than people did, say 100 years ago. The average life expectancy then was about 45 years.

Nowadays, as a teenager, you can expect to live to the age of 77 years if you are a girl, or to 72 years if you are a boy.

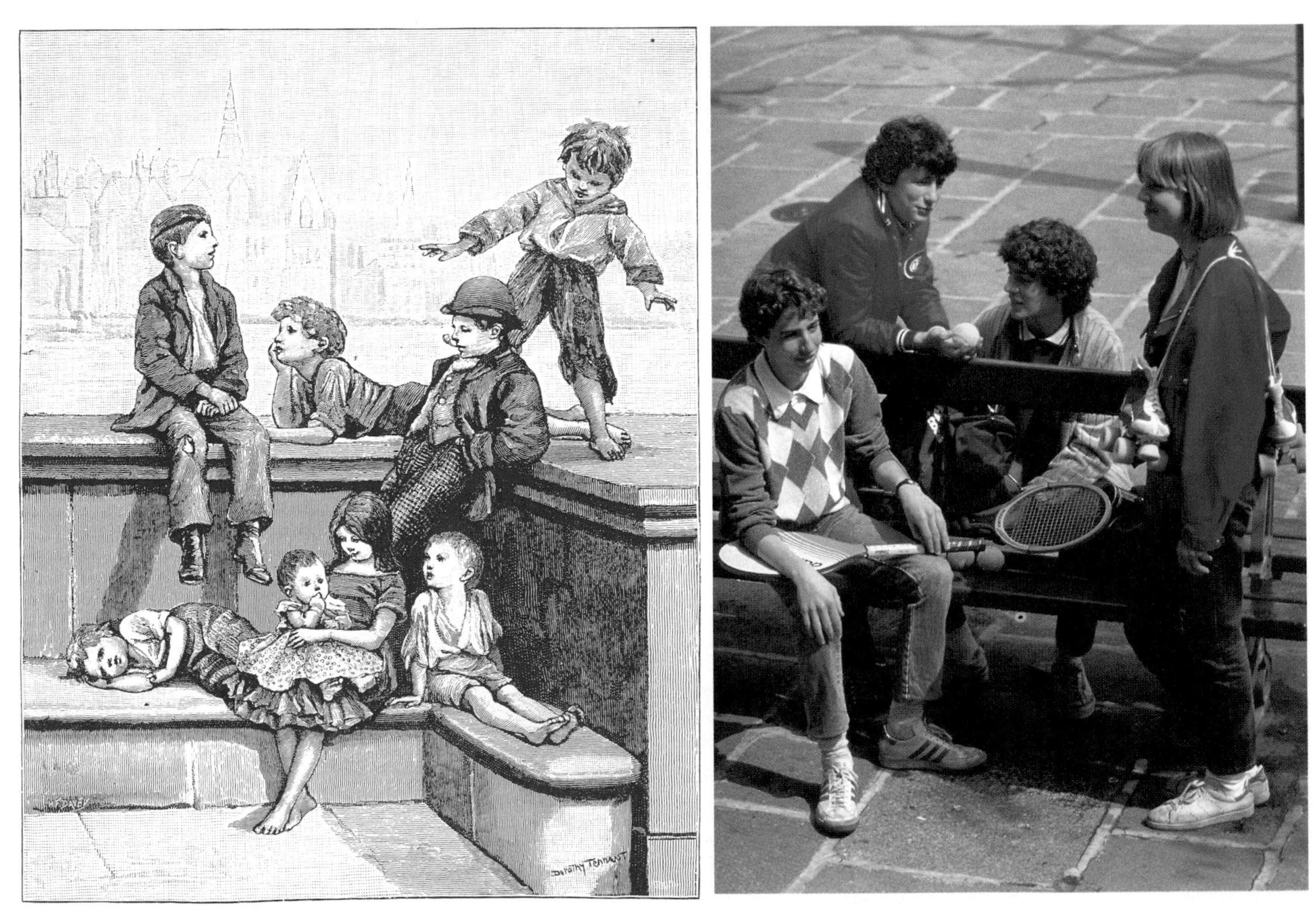

Average life expectancy: 45 years.

Average life expectancy: 75 years.

17.1 Diet and exercise

Starting off

The food you eat provides your body with stores of chemical energy. Most of this is used to provide energy for four things :

- to keep your body working (breathing, heartbeat, and so on)
- for growing new cells in your body
- to keep your body at the right temperature
- to enable your muscles to do work.

Keeping a healthy body weight is a matter of balancing energy you take in as food, with the energy you use up.

If you eat more energy than you use, the extra energy is stored as fat in your body.

Exercise helps you to use up energy. It helps you to keep fit by increasing the body's **metabolism**. The chemical reactions in your body speed up during exercise and for a short while afterwards.

You can keep fit by eating reasonable amounts of sensible foods, and by exercising regularly.

Most of the energy you take in comes from **carbohydrates** and **fats**. These are both substances that contain complicated molecules built up of carbon, hydrogen and oxygen atoms.

Carbohydrates include sugars, and the starches from bread, potatoes, beans, breakfast cereals, pasta and rice. The cellulose fibre from the cell walls of these plant foods is not actually digested. It passes through the body without being absorbed. However, a high fibre diet seems to be important in preventing diseases of the intestine.

Fats include butter, margarine and cooking oils. They are also found in cheese, sausages, beefburgers, chips, crisps and nuts. Researchers have found that too much animal fat, (called **saturated** fat) seems to increase the risk of heart disease. The vegetable fats (**unsaturated** fats) seem to be healthier.

The muscles, tendons and cartilage in your body are made up of **proteins**. So are your blood cells, and your hormones, the chemical messengers in your body. Food containing protein is important so that new cells can be produced.

Protein-rich foods include meat, eggs, fish, soya beans, lentils and peanuts.

Foods rich in

Carbohydrate

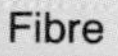

Fibre

Fat

Protein

1 Which of the following are starchy foods :
cornflakes, chicken, toast, potato crisps, lettuce, baked beans ? ▲

2 Which of the following foods are high in protein :
lettuce, beef, eggs, potatoes, chicken, apples? ▲

3 Make a list of the food you eat in a typical day. For each food, note the types of nutrients - carbohydrate, fat, protein - that it contains. (Food labels may help you to decide.)

Did you know ?

- You need to run for about 20 minutes to use up the energy you get from one slice of bread and butter.

17.1 Balancing your diet

Going further

The chemical reactions that take place in your body need minerals and vitamins to control them. **Vitamin A**, for example, helps to prevent infection. **Vitamin C** helps the body to absorb **iron**, which forms an important part of your red blood cells. **Vitamin D** helps the body to use **calcium** and **phosphorus** to make teeth and bone tissue.

Most of our vitamins come from fruit and vegetables.

Iron, calcium and phosphorus are minerals. Although they are essential for good health, like vitamins, they are only required in small amounts.

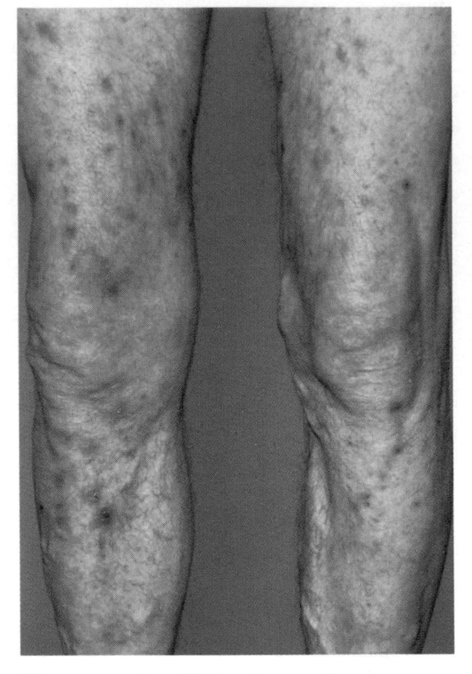

This person is suffering from a disease called scurvy, caused by a lack of Vitamin C.

Water

Don't forget the **water**! Water is needed as the solvent to enable all these chemical reactions to happen. You can live for several weeks without food (although it would be unpleasant!) You would die after only a few days without water.

Getting the right balance

The latest report from the World Health Organisation gives suggestions for a healthy diet. Up to 70% of our energy intake should be in the form of fruit, vegetables, pulses (for example beans and lentils) and food made from grains (bread and cereals for example). This would provide for most of our energy and control needs. A further 15-30% of our intake could be fat, with vegetable fats preferable to animal fats.

Protein should account for 10-15% of our intake.

Added sugar (in the form of fizzy drinks, chocolate bars, biscuits and sweets, for example) is *not essential* for a healthy diet. It should *not* make up more than 10% of our total intake. The *absolute maximum* of added sugar for one day would be, say, one can of cola and one slice of cake.

Malnutrition People tend to think that malnutrition only occurs in under-developed counties where there is not *enough* to eat. But not getting the *right balance* of nutrients can be just as important. In this country for example, cases of malnutrition can occur where people eat only a narrow range of packaged, convenience foods.

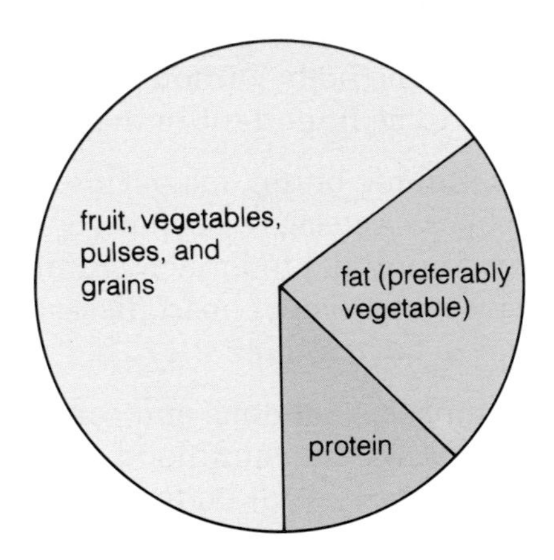

Recommended daily diet.

1 a) What nutrients will be missing if your diet consists mainly of beefburgers, chips, sweets and fizzy drinks?
b) What else would you need to eat to get a better balance? ▲

2 For the foods you eat in a typical day, add the types of nutrients - mineral, vitamin, water - that they contain. Are you eating a balanced diet?

3 **Try to find out** what disease develops in children due to a lack of the mineral calcium. Name two foods which could prevent this.

4 **Try to find out** what disease is caused by a lack of the mineral, iron. What foods might prevent this?

5 There are several distinct forms of vitamin B. **Try to find out** which foods contain them, and the different jobs they do.

Did you know ?

- Of all the food you eat each day, the total mass of vitamins you need is only about 30 mg. That's about the same mass as a single grain of rice!
- Over-cooking vegetables can remove all the vitamin C in them. (Vitamin C is very soluble in water.)

17.1 Food additives

For the enthusiast

Agar, from seaweed, is otherwise known as E406. It is added to many foods to improve texture.

When is a potato crisp not a potato crisp? This isn't a riddle, but if you look on the packet you might seem to be getting more than you bargained for! One brand includes: dried potato, vegetable oil, starch, salt, emulsifier (E471), and flavour enhancer (621).

Many food manufacturers add chemical substances to foods. Some of these **additives** are used as preservatives, some are used to improve the natural flavour of the product. Others can be used to prevent oxidation, to adjust acidity, to alter thickness, to give colour or to add sweetness.

In 1984, new food labelling regulations made it necessary for manufacturers to list the ingredients on food products. Most additives now have an 'E' number to show that they have been approved for food use.

Some of these 'E' additives are harmless and do a valuable job. Some are used for less vital reasons, such as colourings to improve a food's appearance. Some additives have actually been shown to be harmful to some people.

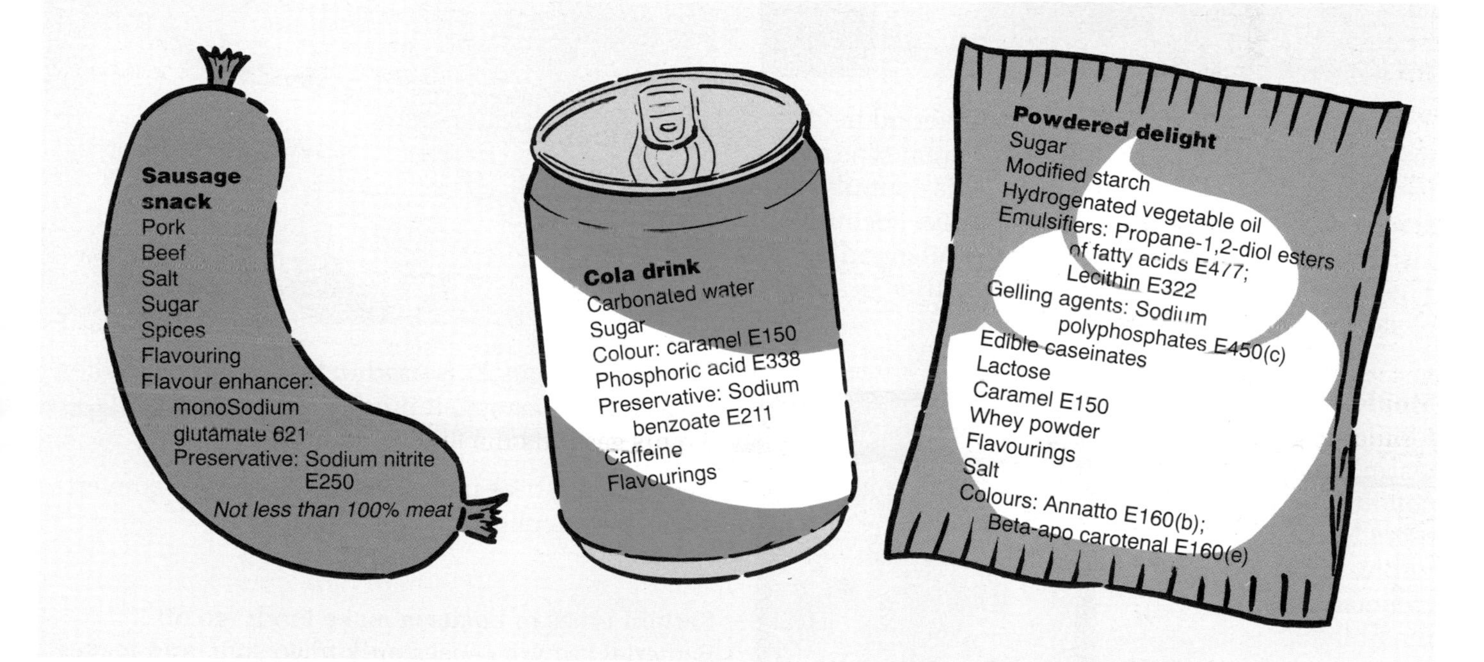

1 What sort of substances are given an 'E' number? ▲
2 Why do you think manufacturers add food colourings to some of their products?
3 Try to find out what some of the additives are in the pictures above. Why are they added?
4 Do any things surprise you about the 'Sausage snack' label? ▲
5 Write a letter to your school newspaper explaining either:
   why you think additives are a good thing,
   why you think additives are a bad thing, or
   why you think some additives should be allowed, but others shouldn't.

Did you know?

- The food colouring E120 (cochineal) is made from the dried bodies of pregnant scale insects which live on cactus plants in Central America.

17.2 Microbes and food

Starting off

Fungi and **bacteria** are two types of microbe that are very important to healthy eating. You need a microscope to see these tiny living **micro-organisms.** Yet in spite of being so small, they can have a great effect on you, and on what you eat.

These microbes can be 'goodies' or 'baddies'! The 'goodies' are important in the making of certain foods. The 'baddies' can make foods go bad, as well as causing disease.

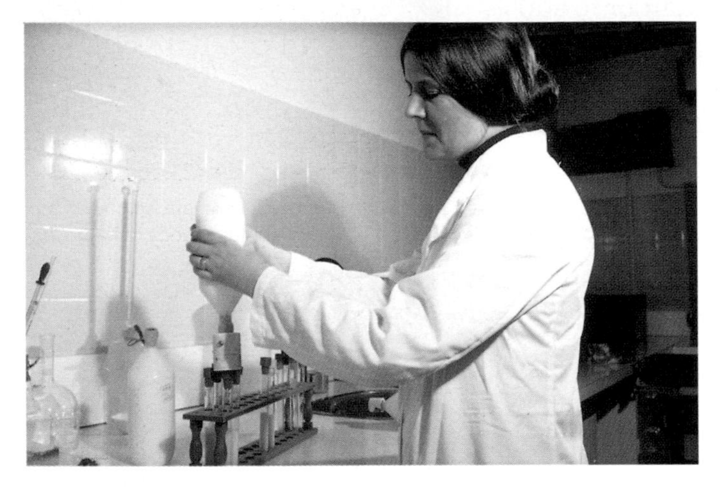

To make cheese the bacteria normally found in milk are first killed by boiling. A different type of bacteria is then added to the milk. This controls the acidity of the milk, and alters its flavour. Yoghurt is also produced by adding bacteria to boiled milk. The process was first used in hot countries as a way of stopping milk going sour.

Yeast is a fungus. It is used in bread-making. When yeast feeds on sugar, it produces carbon dioxide. This gas puts the 'bubbles' in bread.

In wine- and beer-making, yeast is used to convert sugar into alcohol.

Moulds are fungi. Moulds grow in warm, damp conditions. They often have an unpleasant taste, and can make foods inedible.

Certain types of **bacteria** make foods 'go off'. Bacterial growth causes milk to go sour, and makes meat and fish go rotten.

1 List the main types of microbe and their effects. ▲
2 Make a list of foods which use fungi in their production. Try to find out others not mentioned here.
3 Name a fungus you can eat which isn't a microbe.
4 What would be an alternative way of producing the bubbles in bread?
5 **Try to find out** other types of food which use bacteria in their production.

Did you know ?

- The blue veins in Stilton and Danish blue cheeses are made by encouraging the growth of moulds.
- A pot of yoghurt can contain more bacteria than there are people on earth.

17.2 Preserving food

Going further

There are many ways which can be used to **preserve** foods. Each way of keeping foods longer depends on either:

- killing the microbes, or
- slowing down the rate at which the microbes grow, or
- preventing microbes getting to the food.

Microbes grow quickly in warm conditions. Freezing or refrigerating food slows down the growth and the food keeps fresh longer. This does not, however, kill the microbes.

Microbes grow quickly in damp conditions. Certain foods are preserved by drying. Again, this does not kill the microbes.

Heat kills microbes. Cooked meat and fish keep for longer than raw. Normal cooking, though, does not kill *all* bacteria. Cooked foods need to be refrigerated to slow the growth of remaining bacteria.

Canning and bottling are used to preserve foods for long periods. The food is heated at around 120°C to kill bacteria. It is then sealed in the container to prevent the entry of new microbes.

Jamming, pickling and salting are used to preserve foods. Strong solutions of sugar, vinegar or salt are used. This kills the microbes by drawing the water from their cells.

A new (but controversial) method of food preservation is to use radioactivity. Food is prepacked and then irradiated. This kills the bacteria without affecting the cells of the food.

1 You are writing a recipe for bottling fruit. Which stages would you stress to make sure that the fruit would keep well? ▲
2 Why can dried fruits be kept for much longer than fresh fruit? ▲
3 'Steak tartare' (minced steak mixed with egg yolk) is eaten raw.
   a) What are the dangers of eating raw meat and eggs?
   b) Which of the above methods could be used to kill the bacteria in the meat and egg? (Heating isn't allowed!) ▲
4 Chickens and turkey contain salmonella bacteria.
   a) Do frozen chickens and turkeys contain these bacteria? ▲
   b) Why is it important to let the bird thaw out completely before cooking it?
5 Why is it dangerous to thaw frozen meat and then refreeze it without first cooking it?

Did you know?

- Louis Pasteur realised that microbes caused food to go bad. He showed wine producers how to stop wine going sour by heating it to 60°C for a short time.
- This process became known as **pasteurisation** and is used nowadays to help milk to keep longer.

17.2 Pasteur and bacteria

For the enthusiast

Food for an army. A French chef, Nicolas Appert, seems to have been one of the pioneers in developing food canning as a preserving method. In 1810, he won a French government prize of 12 000 francs for developing a method of storing food. The food was needed to feed the army. It had to be easily portable, and it had to keep fresh for several years. His method depended on heating food in a container to *remove* the air and then sealing it to *keep out* the air. He thought it was the *air* that made food go off. He discovered his method purely by experimenting. The scientific reasoning behind his success was not discovered for another 50 years.

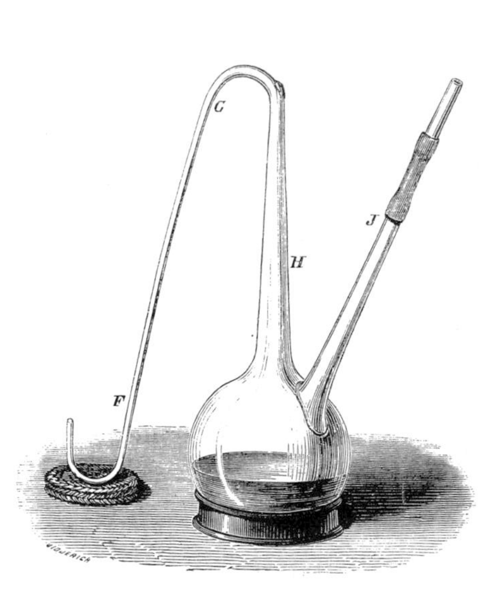

Louis Pasteur

In the 1860s the Frenchman Louis Pasteur reasoned that it was not the air itself that caused decay, but microbes in the air, possibly carried on dust particles. He carried out a series of experiments to test his theory. He used the elegant swan-necked flasks shown in the picture.

Some students are trying to repeat one of Pasteur's experiments. They are using the simpler apparatus shown in the drawings below.

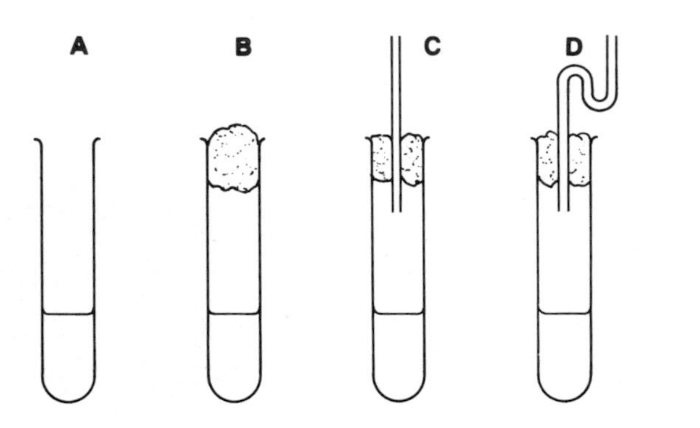

1 They poured nutrient broth (a kind of clear gravy) into four test tubes.
2 They put various types of stopper in three of the test tubes, as shown.
3 They placed the four tubes, complete, in a pressure cooker and boiled them under pressure for 20 minutes. This killed all the bacteria present in the broth, in the test tubes and on the glass tubes and cotton wool.
4 They then let the test tubes cool, and left them out in the lab for the next few weeks.

The students wanted to find out if the broth goes off because of the air itself, or because of bacteria in the air dropping into the broth .

As long as the broth stays fresh, it will remain clear.

If the broth goes off, it will start to go cloudy.

1 Tube A can let in air *and* bacteria. Do you think the broth will go cloudy?
2 Tube B keeps out air *and* bacteria. Do you think the broth will go cloudy?
3 Tube C has a *small* opening which lets in air and through which bacteria could fall onto the broth. Do you think the broth in tube A or tube C will go cloudy first?
4 Tube D can let in air, but any falling bacteria would be caught on the side of the bent tube, and wouldn't fall into the broth.
  a) If air was responsible, would the broth in this tube go cloudy?
  b) If bacteria were responsible, would the broth in this tube go cloudy?
5 In the experiment, tube A went cloudy in a few days, followed shortly by tube C. Tubes B and D were still clear after two weeks. What do you think was their conclusion?

Did you know?

- Pasteur went on to use his knowledge of microbes in the fight against disease. His most famous work is in connection with fighting rabies.

17.3 Diseases

Starting off

There are several ways in which the body can go wrong. Sometimes, in cancer for instance, cells in the body behave in an abnormal way. Other disorders, for example beriberi or rickets, are due to **malnutrition**. Certain conditions, for example heart disease, might be due to our **lifestyle**: eating too much saturated fat, not taking enough exercise, smoking or drinking alcohol.

Another group of diseases is caused by **infection**.

A few infectious diseases are caused by **parasites**. Roundworms, threadworms and tapeworms are parasites which can live in various parts of the human intestine. The eggs get into the body on uncooked meat, or on food which has been touched with soily hands. The eggs then hatch inside the body, and the worms live on *your* food.

The majority of infectious diseases are transmitted by four types of microbe. Most of these diseases are caused by **bacteria** and **viruses**. A few are caused by **fungi** and **protozoa**.

The head and neck of a beef tapeworm. It is a parasite of the human intestine, and can be up to 10m long. (Magnified x15)

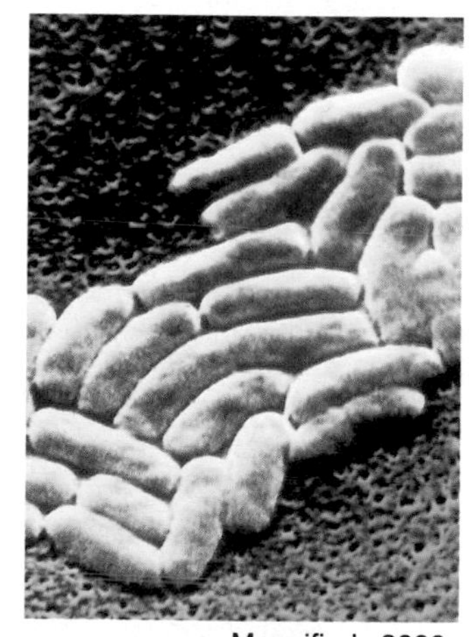

Magnified x3000

Bacteria

Diarrhoea, tetanus, blood poisoning, cholera and leprosy are caused by bacteria.

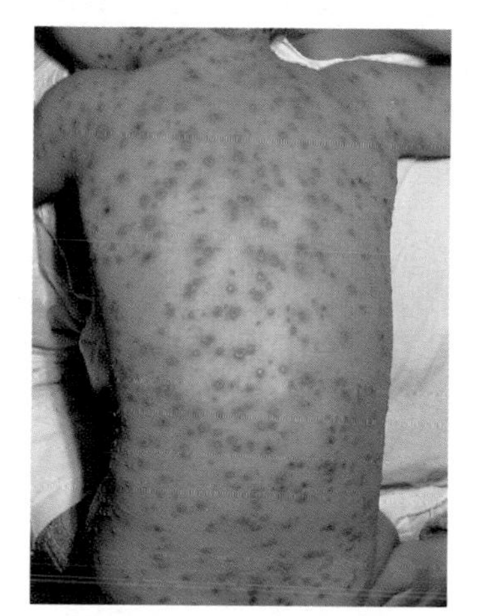

Viruses

Diseases caused by a virus include German measles, flu, chicken pox, mumps, smallpox, cold sores, glandular fever, colds, rabies and AIDS.

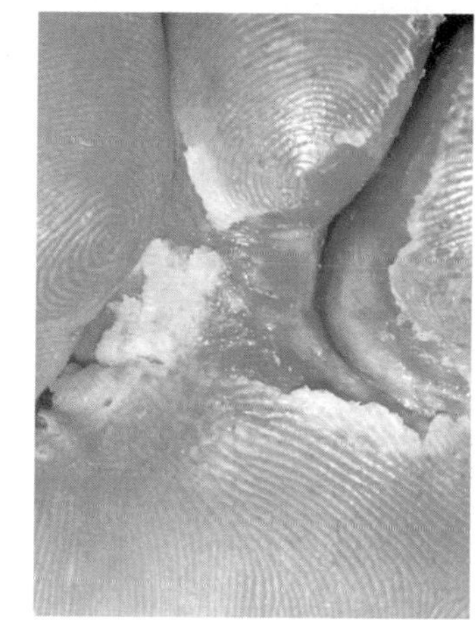

Fungi

'Athlete's foot' is a fungus which grows in the warm, moist areas between your toes.

Protozoa

Malaria is caused by a protozoan that lives in human red blood cells. It is transmitted by infected female mosquitoes.

1. Use a dictionary to find the definition of a parasite.
2. Make a list of things which can cause disease. ▲
3. Why is between your toes a good place for a fungus to grow? ▲
4. Make a list of any infectious diseases you had when you were little. Against each disease write the type of microbe which might have caused it.

Did you know?

- Microbes are very small in size:
  fungi: about 0.01 of a millimetre
  bacteria: about 0.001 of a millimetre
  viruses: about 0.0001 of a millimetre
- Bacteria multiply by dividing! In the right conditions, each bacterium splits into two every 20 minutes. After 7 hours, a single bacterium would have produced over 2 million bacteria!

17.3 Prevention 1

Going further 1

Hygiene

This children's nursery rhyme is supposed to have originated in London in 1665 during the Great Plague. Sneezing was one of the symptoms of the Plague which swept through the city killing 60 000 of the 450 000 population. People carried posies (or bunches) of sweet smelling flowers to hide the smells of the street. They thought it would prevent them catching the disease. It didn't, and they 'all fell down' dead.

Ring-a-ring of roses,
A pocket full of posies,
A tishoo! A tishoo!
We all fall down!

Bubonic plague was passed from rats to people by fleas.

Three things would have prevented the spread of disease:

a much better standard of hygiene,
uncontaminated drinking water, and
better sewage disposal.

Joseph Lister and antiseptics

The British surgeon Joseph Lister (1827-1912) noticed that patients' wounds often went bad after operations. He carried out a survey of people with fractured bones. He observed that simple fractures always healed well. But compound fractures, where the bone pierced the skin and produced a wound open to the air, often went bad. He came to the conclusion that microbes in the air were infecting the wounds.

Lister had read of Pasteur's work on killing microbes. In 1865 he carried out a trial operation. He washed all scalpels and bandages in carbolic acid. The wound, too, was sprayed. This **antiseptic** successfully killed the microbes and prevented infection.

Lister followed this procedure with all his operations. Wounds were re-dressed with antiseptic bandages. Within two years infection in the hospital was dramatically reduced.

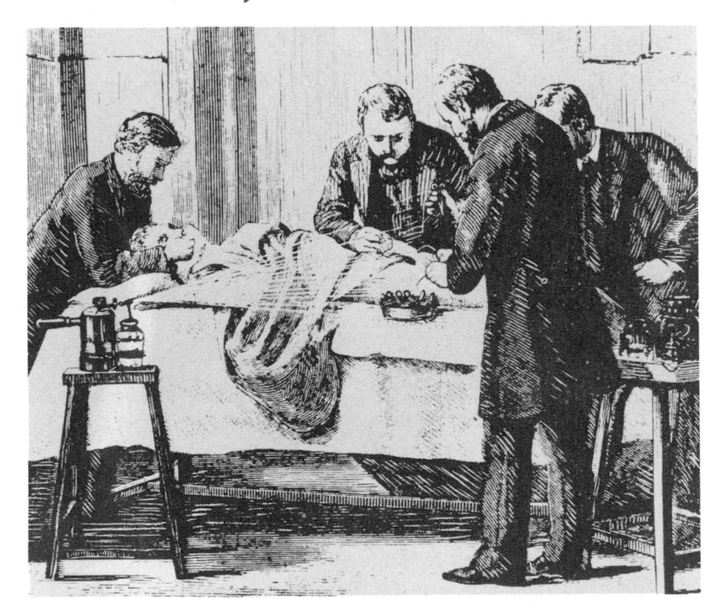

Spraying a wound with antiseptic.

Hospitals now **sterilise** their instruments to kill microbes on them. High temperature, high pressure, autoclaves can be used. Dressings can be packaged and treated with radioactivity to kill microbes.

Did you know?

- An uncovered sneeze can spread the common cold virus up to 10 metres.

1. What three things could have helped prevent the spread of the Great Plague? ▲
2. People held sweet flowers to their noses to try to prevent disease. They thought it was the smelly air that carried the plague. Why were they wrong? ▲
3. Why do surgical staff wear masks in the operating theatre?
4. *T.C.P.* and *Dettol* are antiseptics. How do they work? ▲
5. **Try to find out** about modern sterilisation techniques.

17.3 Prevention 2

Inoculation is a method of making the body immune to a disease by introducing a mild form of the disease into it.

Inoculation to prevent a disease in animals was described by Rider Haggard in 1885 in his book 'King Solomon's Mines'. (Not all science has to be in science books!)

The panel on the right summarises his description.

"These oxen had all been inoculated against 'lung sick', which is a dreadful form of pneumonia. This is done by cutting a slit in the tail of an ox, and binding in a piece of the diseased lung of an animal which has died of the sickness. The result is that the ox sickens, and takes the disease in a mild form. This causes its tail to drop off, but the animal is protected from further attacks."

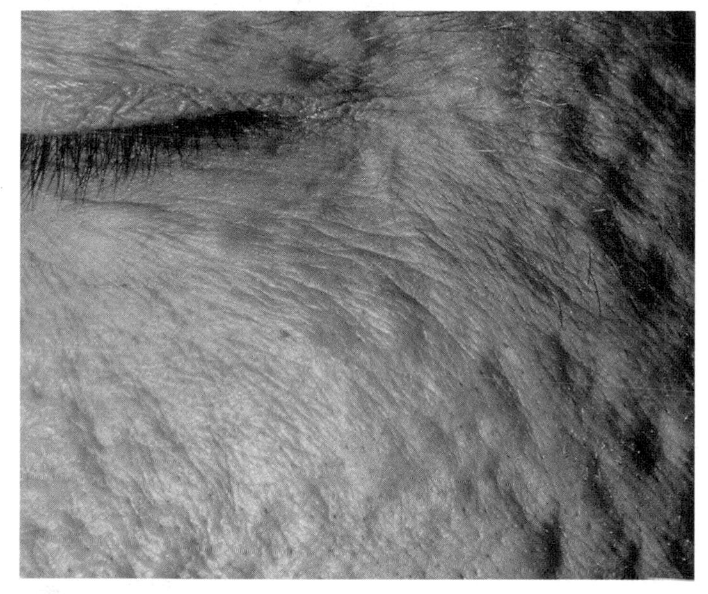

Smallpox leaves permanent scars on its victims

Edward Jenner and smallpox

About 200 years ago, smallpox was a killer disease. Even people who recovered from the disease could be scarred for life by the blisters that formed on the skin.

At that time country people believed that farmworkers who caught cowpox from their cattle were safe from smallpox. A Gloucestershire doctor called Edward Jenner made a careful survey and observed that this appeared to be true.

Through research he showed that cowpox was a mild form of smallpox. He found that, if he injected someone with fluid from a person suffering from cowpox, they would not catch smallpox. They became **immune** to it and were safe even if they then came into contact with someone who had got smallpox. This injected fluid is called a **vaccine**.

Jonas Salk and polio

Poliomyelitis (polio for short) is a virus disease which attacks the spinal cord. The disease destroys nerve cells and its victims become paralysed. Around 1950 it was widespread throughout Europe and America. In 1953, Salk developed a vaccine to prevent polio. He obtained samples of the virus and killed them with a formaldehyde solution. Injecting the dead virus into people gave them immunity from the disease. Polio is now almost unknown in developed countries.

Inoculation is used nowadays to prevent a wide range of diseases. These include whooping cough, diphtheria, tetanus, polio, measles, German measles, smallpox and influenza.

1 How does inoculation work? ▲
2 What stages did Jenner go through in his research to prevent people catching smallpox? ▲
3 Inoculation of infants has wiped out many diseases. What would be the result if all parents decided *not* to have their children inoculated?
4 **Try to find out** why Jenner called his method of inoculation 'vaccination'.

Did you know?

- Many centuries ago, the Chinese discovered that they could help to stop the spread of smallpox. They inhaled the powdered scabs from smallpox victims.

17.3 Cure

For the enthusiast

The body's own defence system

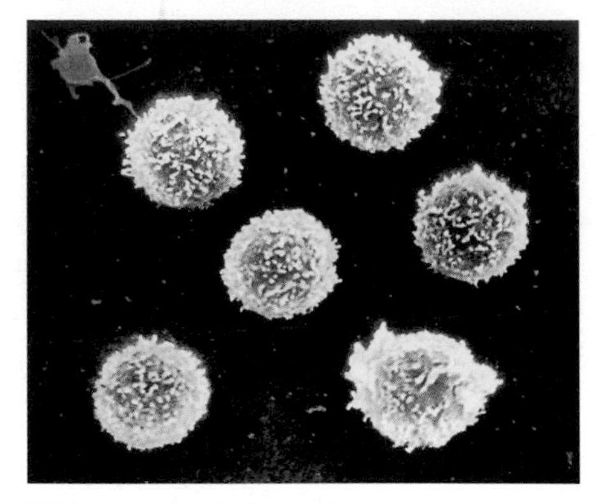
White blood cells (x2800).

Your body has a good defence system against microbes. You have your own protective wrapping - your skin!

For alien microbes which do get in, your body is able to recognise them as 'enemies'. A defence force comes into action. Certain types of **white blood cell** surround the invaders and make them harmless. Other white blood cells are able to make **antibodies.** Each type of antibody is designed to attack and kill a certain type of microbe.

After one infection, these white blood cells 'learn' to make that type of antibody. They come into action more quickly next time you are infected. You are then **immune** to that disease.

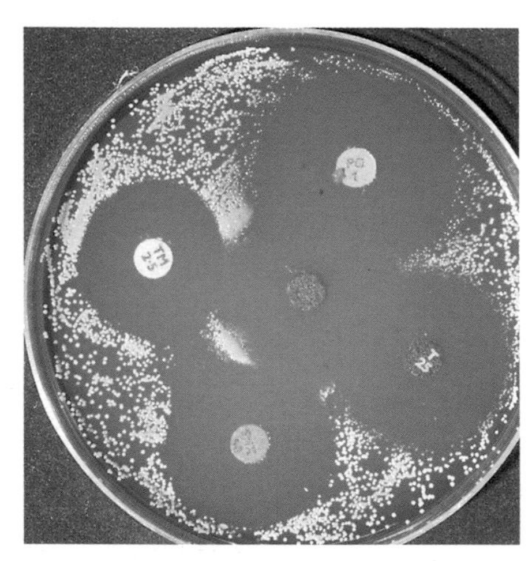
The pink discs are antibiotics. As they dissolve they kill off the bacteria surrounding them.

Drugs

Drugs are chemicals that can be used to relieve symptoms or to control illnesses. The most well known is probably 'aspirin'. This is used as a pain killer, and also to lower a patient's temperature.

Some drugs act on the disease itself by killing or controlling the spread of microbes. **Antibiotics** are substances which can kill or control **bacteria.**

The first antibiotic was discovered by accident! In 1928 Alexander Fleming, a Scottish scientist, was studying the bacteria which stopped soldiers' wounds from healing. One of his culture plates had been left open, and a mould had started growing on it. Before throwing the plate away, he examined it. The mould, called *Penicillium notatum*, had stopped the bacteria from spreading. It was producing a substance which killed the bacteria. From this discovery, Howard Florey pioneered the use of the substance - **penicillin** - to treat patients.

Radioactivity

The energy from radioactive substances can kill body cells. This is why radioactivity is potentially dangerous. But this ability to destroy human cells can be of value. If the radiation can be concentrated in one small area, it can be used to kill the damaged cells there.

One method of destroying cancerous cells is to direct a beam of **gamma radiation** on them. The damaged cells in arthritic joints can be killed by injecting a radioactive liquid into the joint.

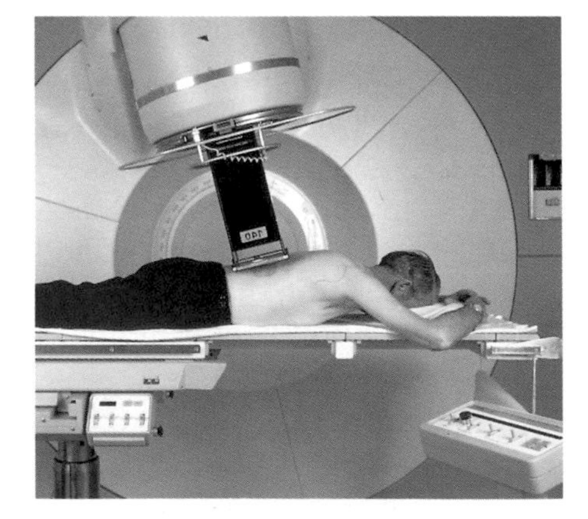

1 Give two ways in which microbes can enter your body to cause disease.
2 How do white blood cells help protect the body from disease? ▲
3 In AIDS (auto immune deficiency syndrome) a virus causes the body's **immune system** to fail. How would this affect a person's ability to keep free from other diseases?
4 Why can't antibiotics be used to cure the common cold?

Did you know?

- Many hospital instruments are now sterilised by prepacking them, and passing them through a high level of radioactivity.

17.4 Nature or nurture?

Starting off

Some aspects of your appearance are decided before your birth. Certain information is passed to your body cells on **genes** from the egg and sperm cells of your parents. Your eye colour and what sex you are, for example, are decided completely by your genes. Other aspects, such as your height and weight will be influenced by external factors such as the food you eat.

So what you look like, and the way you develop, are influenced by both **genetic** and **environmental** factors.

Tongue rolling

Try to curl up the sides of your tongue so that they meet in the middle. Can you do it ? Some people can and some people can't, even with practice. The ability to roll your tongue is controlled only by genetic factors.

Body building

This woman has spent a lot of time in the gym to look completely different!

However, if she has children, they won't be muscle-bound because this aspect of her appearance is caused by environmental factors.

Twins

Identical twins have the same genetic make-up. Any differences which become apparent as they grow up are likely to be due to environmental factors.

1 What factors can affect your appearance? ▲
2 Make a list of things you think might be genetically controlled. (Use the examples given to start you off.) ▲
3 There are many environmental differences between a developed country and a developing country. In what ways might these affect the way a child grows?

Did you know ?

- 
  Hydrangeas show a good example of the way the environment can alter the characteristics of an organism. If they are grown in acid soils, their flowers are blue. In neutral or alkaline soils they are pink or white.

17.4 From parent to child 1

Going further 1

Passing on the message: heredity

The instructions for designing a new baby are contained in the egg cell (**ovum**) and sperm cell of its parents.

Fine, thread-like structures called **chromosomes** in the cell nucleus carry these instructions. Most human cells contain 23 *pairs* of chromosomes. However, when sperm and egg cells are being made, these pairs separate so that sperm and egg cells each contain only 23 *single* chromosomes. When the sperm fertilizes the egg a **zygote** is formed which has a full complement of 23 *pairs* of chromosomes. Twenty two of these pairs match each other, but the remaining pair can differ from each other. These are the **sex chromosomes** (X and Y) which control whether the baby will be a girl or a boy. In females both chromosomes are the same (X and X). In males there is one X and one Y chromosome. The diagram below shows how the chromosomes may combine.

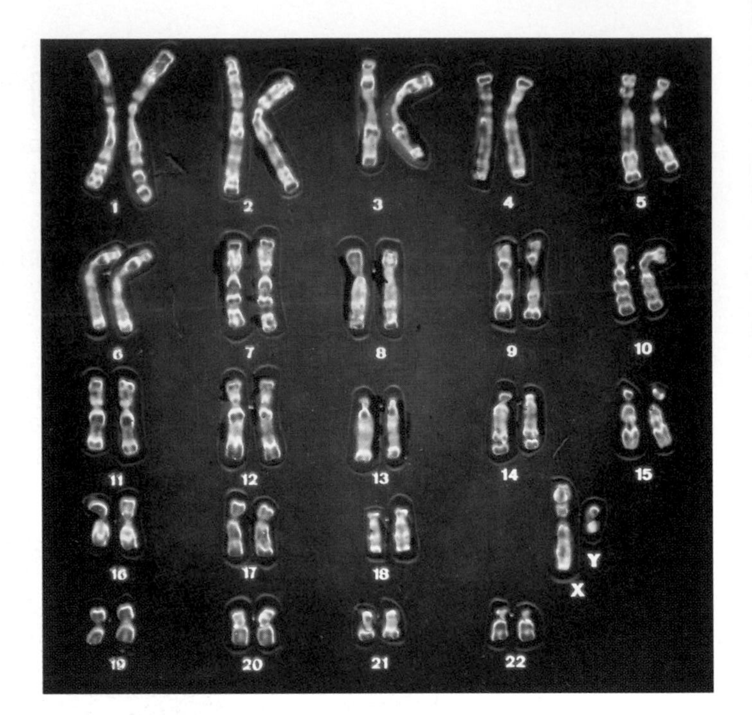

A full set of human chromosomes.

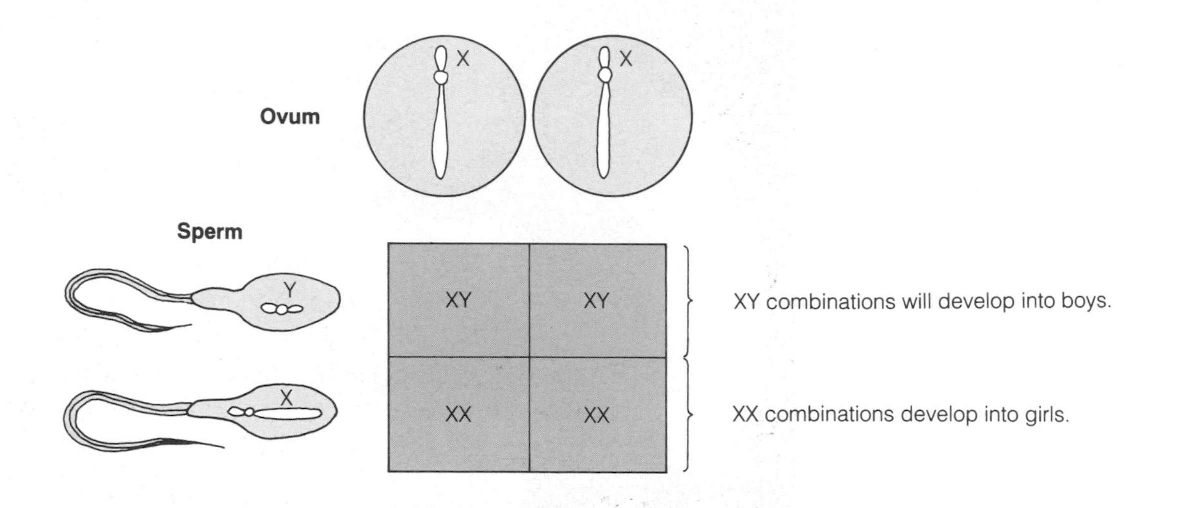

Every chromosome contains up to 10 000 genes. Genes are made from a chemical called **DNA** (short for **d**eoxyribo**n**ucleic **a**cid). Each gene carries a specific instruction for the 'design' of the baby. Some genes are more **dominant** than others: they pass on 'stronger' messages. Weaker genes are called **recessive**. Your mix of genes will decide whether you are tall or short, fair or dark, blue- or brown-eyed, left- or right-handed.

1 Describe how genetic information is passed from parents to children. ▲
2 What decides whether a baby is a boy or a girl? ▲
3 What are genes made of? ▲
4 What is a dominant gene? ▲

Did you know?

- Unlike humans and many other animals, butterflies, moths, birds and some fish have XX males and XY females.

17.4 From parent to child 2

Along with the instructions for life, genes sometimes carry disorders from parent to child. These disorders are known as **hereditary diseases**.

Queen Victoria was a **carrier** of the gene for **haemophilia**, a disease in which blood fails to clot. As clotting is vital to stop bleeding from cuts, any small wound can be fatal to a haemophilia sufferer.

Queen Victoria did not suffer herself from haemophilia, as she also had the non-haemophilia gene which is dominant over the haemophilia gene. However, two of her sons were sufferers.

Thalassaemia, a blood disease, and **Duchenne muscular dystrophy** are other examples of hereditary diseases.

In the womb

A baby may be harmed before birth by substances passing to it from the mother while it is in the womb.

The placenta acts as a filter between the mother's blood system and the baby's. But some cells are small enough to pass through the filter. If a pregnant woman catches German measles, her baby may be born deformed or mentally retarded. It may have a damaged heart, or cataracts on the eyes leading to blindness.

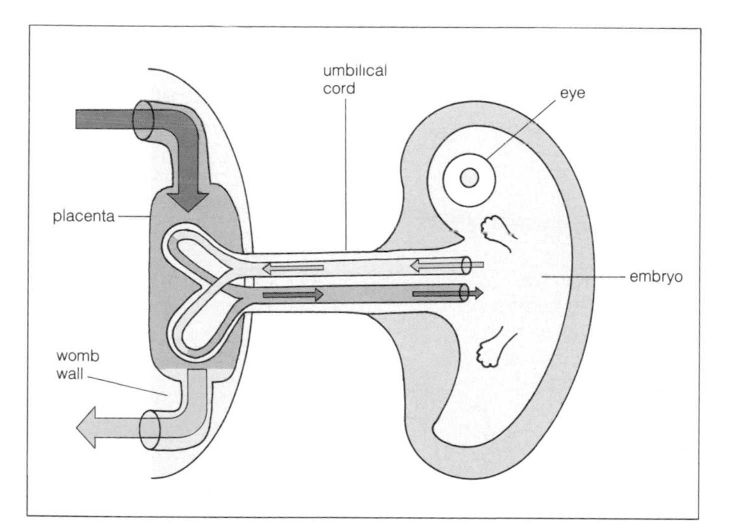

The microbes of some sexually transmitted diseases may also be passed on. The baby may be born deformed, deaf or blind, as well as having the disease itself.

Some chemical substances can be passed from mother to baby. Pregnant women with a high alcohol intake may give birth to babies with physical and mental disorders. The chemicals from some medicinal drugs may also affect the growing **foetus**. The foetus obtains oxygen across the placenta from the mother's blood. If the mother smokes, her blood carries less oxygen to the foetus, and this may cause her baby to be born sickly and underweight.

1. Describe the ways in which a foetus is affected by substances in the mother's blood. ▲
2. Why is it unwise for a pregnant woman to smoke or drink? ▲
3. Why do girls have a Rubella (German measles) vaccination at around the age of 12 or 13?
4. Why are boys not given this vaccination ?
5. If an older woman has a Rubella vaccination, why might it be dangerous for her to become pregnant in the following couple of months?

Did you know ?

- Dwarfism is a hereditary disorder where the limb bones do not grow. In families with a history of dwarfism, half of their children are likely to inherit the condition.

17.4 Peas and genes

For the enthusiast

For thousands of years people have controlled the breeding of plants and animals to suit their needs. They knew that by choosing the parents carefully, the 'offspring' were likely to have the right features. Sweeter apples, cattle which produced more milk, and better hunting dogs were among the products of this **selective breeding**.

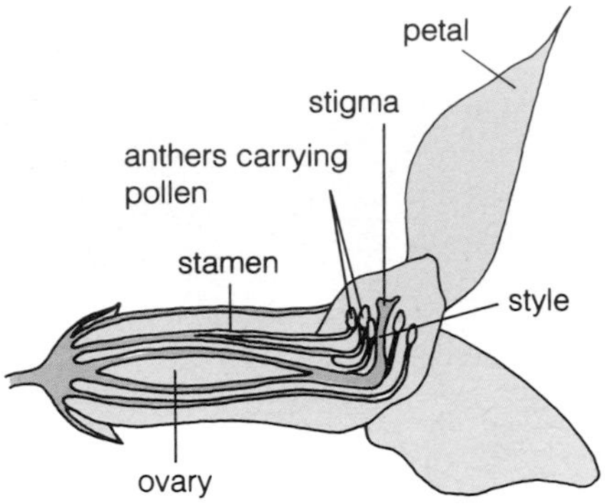

However, although people could control breeding, they did not know for sure what carried the message from generation to generation. Some thought that blood from each parent animal mingled to produce a mix of features in the offspring.

Not until 1865 did the way heredity works become well understood. Gregor Mendel, an Austrian monk, had carried out a series of carefully planned experiments using pea plants. He started off by fertilising pea plants with their own pollen (self-pollination), and then planting the seeds. According to the size of the seedlings, he then sorted his original plants into two 'pure' groups:

- tall plants which always produced tall offspring

  and

- dwarf plants which always produced dwarf offspring.

petal
stigma
anthers carrying pollen
stamen
style
ovary

Using these pure strains of pea plants, Mendel experimented to see what happened when he cross-pollinated them. He first had to remove the anthers from some of the tall and dwarf plants to prevent self pollination. Then he :

- put pollen from tall plants onto the stigmas of dwarf plants

  and

- put pollen from dwarf plants onto the stigmas of tall plants.

When the seeds ripened he planted them and measured the size of the offspring seedlings (or hybrids, as offspring are called when the parents are of different types).

The diagram shows what can be observed about inheritance of one gene controlling one feature (a **monohybrid cross**). **T** represents the tall gene and is dominant, and **t** represents the dwarf gene and is recessive.

In all, he studied over 21 000 plants and analysed the results of breeding, or **crossing**, them. Mendel concluded that inheritance is through individual 'instructions' (now known to be genes) rather than a general blending of features.

Did you know?

- All today's dogs are varieties of the same species. They have been bred for particular purposes over centuries.
- Tigers and lions can mate with each other to produce 'ligers' and 'tigrons'. These hybrids, though, are infertile.

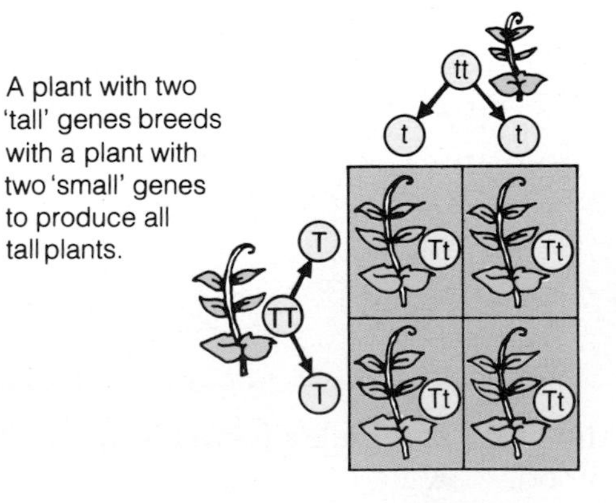

A plant with two 'tall' genes breeds with a plant with two 'small' genes to produce all tall plants.

1. Name four types of specially bred animal and plant.
2. In what ways was Mendel careful to ensure that the results of his experiments could be relied upon? ▲
3. What did Mendel decide about the way in which features are inherited? ▲
4. Why is it useful to know what controls inherited features?
5. Try to predict what sizes of plants are produced if both parents have genes **Tt**. Use a diagram like the one above.

17.5 Human life cycle

For everyone 1

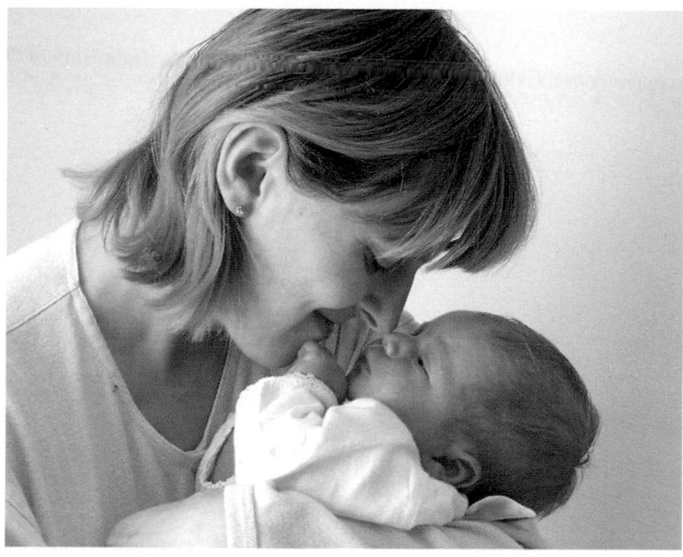

Childhood In the womb a foetus is in warm surroundings, with food and oxygen supplied directly to its blood. At birth, huge changes occur. The new baby has to start breathing, and relies completely on adults for food and warmth. The ideal situation for a baby to grow up in seems to be a loving family with mature parents. The way humans have evolved tries to ensure this is so. But for various reasons (for instance death, or divorce) this sometimes isn't possible. What *is* important is that there are loving adults available. They can help the child to develop into a caring and responsible person.

Adulthood As adolescents mature into adults, another set of changes is likely to begin. Adults may form new family groupings and start having children themselves: the human race continues!

As adults get older, they continue to learn through their experiences at work, in leisure activities, and as parents.

As adults grow into **old age** yet another period of change occurs. Most people retire from work by the time they are 65, and for some this can be a difficult adjustment. Old people may have to cope with disease and disability. Gradually, they are likely to become more dependent on others.

Adolescence is a period of great change:

- There are the **physical** changes (called **puberty**) from being a child towards being an adult. These include the body changes which make it possible to have children.
- **Socially**, adolescents are changing from children dependent on their families, towards being independent adults.
- **Emotional** changes come from two sources:
  - as a result of the chemical changes in the body, and
  - from coping with more, and changing, responsibilities.

1 Make a list of things that a baby needs to help it develop towards maturity. Don't forget obvious ones like food and warmth! ▲
2 What are the main changes which occur in adolescence? ▲
3 List some of the ways in which old people may need help.
4 Humans have evolved so that children are unable to reproduce. What problems would occur if, say, six year olds could have babies?

17.5 Parents and children

For everyone 2

Being a parent carries great responsibilities. As well as the love that all children need, babies are completely dependent on having everything done for them. This takes a great deal of time, effort and money!

On the previous page you may have discussed the problems that were likely to arise if six year olds were able to have babies. Young adults may even find it difficult to cope. They find that trying to earn enough money to run a home and provide all the things that a child needs leaves them very little time with their child.

Adults might feel unable to provide the support needed by a baby, or have other reasons for not wanting to start a family. They may choose to use birth control. There is a range of contraceptives which prevent a woman from becoming pregnant after sexual intercourse, if they are used correctly.

A **condom** is a sheath of thin rubber which fits over the man's penis. A **diaphragm** is a thin rubber cap which fits at the top of the vagina over the entrance to the uterus. Both these contraceptives work by stopping the male sperm cells from reaching the female egg cell. A **coil** works differently. This is a tiny coated metal spring which is fitted into the neck of the uterus. It irritates the lining of the uterus and prevents a fertilised egg from growing there. **Contraceptive pills** work by altering the hormones in the woman's body and preventing the release of eggs each month.

Having sexual intercourse carries the risk of catching a sexually transmitted disease (STD for short). Some STDs such as syphilis, are bacterial. Though these are unpleasant, they can be cured by antibiotics. **AIDS** (**a**cquired **i**mmune **d**eficiency **s**yndrome) is caused by HIV. This stands for **h**uman **i**mmunodeficiency **v**irus. The virus can be passed from one person to another if blood or semen gets into the other person's body tissues. Wearing a condom can help prevent infection by HIV because it provides a barrier between sexual partners. It may take several years for someone with HIV to develop AIDS. AIDS causes a breakdown of the body's **immune** system. The body becomes unable to fight any disease. At present there is no known cure. AIDS is a killer.

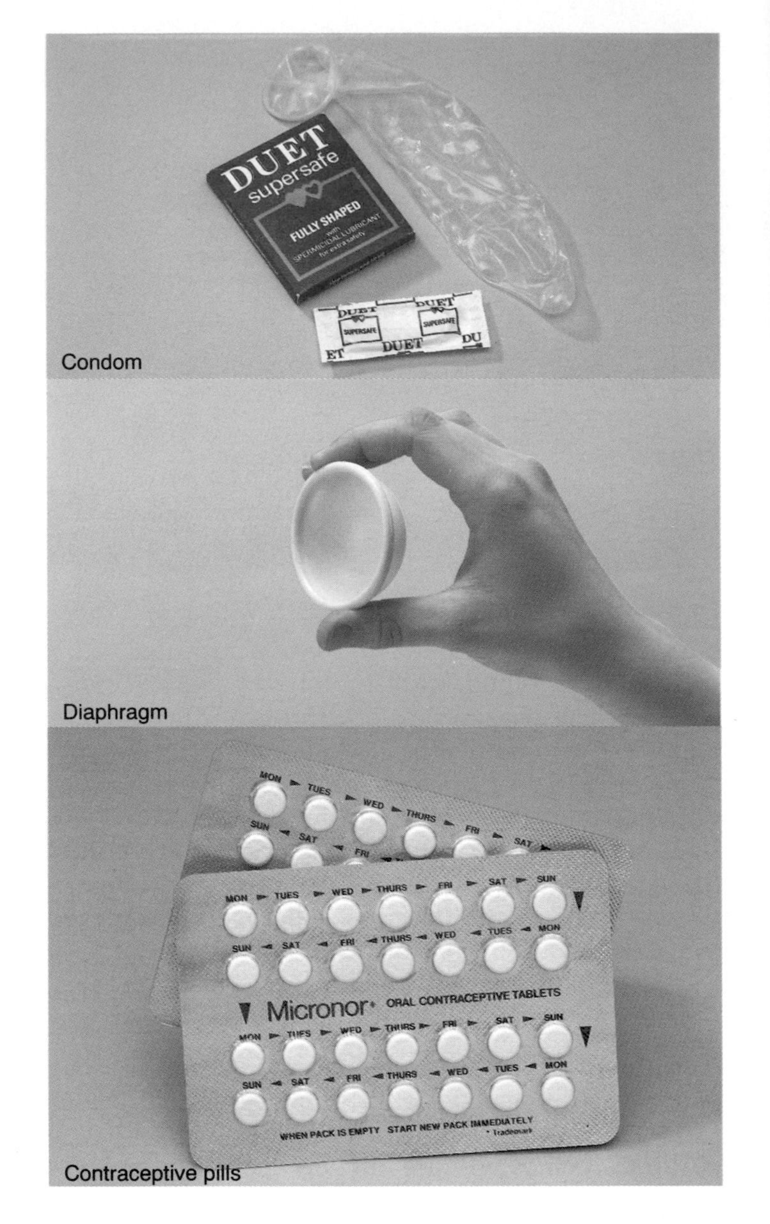

Condom

Diaphragm

Contraceptive pills

1 What changes occur at puberty in a) boys, and b) girls, which enable them to be able to reproduce?
2 What are the *dangers* of having sexual intercourse? ▲
3 What protection against these dangers is there? ▲

Did you know?

- No contraceptive is 100 % safe.
- The more sexual partners a person has, the greater is the risk of AIDS.
- Drug addicts may also become infected with HIV by injecting with second-hand infected needles.

17.5 Your health at risk

For everyone 3

Some people are unlucky and may have a serious disease through no fault of their own. Others make it more likely that they will become ill by smoking, or drinking too much alcohol. A few people risk their health even more by taking harmful illegal drugs.

Smoking Nearly 100 000 people a year die prematurely from smoking-related diseases such as cancer, heart disease and bronchitis. Young smokers who then give up, greatly reduce their chances of developing these diseases.

Alcohol Beer, lager, cider, wine and other alcoholic drinks contain ethanol. This is absorbed into the bloodstream and carried to the brain. A little ethanol makes people relaxed, but too much can cause them to lose control over their speech, movements and emotions. Over 20 000 people in Britain are killed or injured each year in crashes caused by drivers who have drunk alcohol. High levels of ethanol in the blood can cause unconsciousness and even death. Your liver treats ethanol as a poison and works hard to break it down. If a person drinks heavily for some time they risk liver and heart disease, brain damage and cancer of the digestive system.

Solvents Sniffing solvents, glue or paint is extremely dangerous because it is hard to know just how much will produce an overdose. The inhaled chemicals are absorbed into the blood and carried to the brain. Long-term use can cause brain damage, but the major risk is from having accidents, or of choking to death while unconscious.

Illegal drugs Cocaine, crack, ecstasy, LSD, and cannabis are illegal drugs. Heroin, tranquillisers, barbiturates and amphetamines are medicinal drugs available on prescription, but they are illegal if they are not prescribed to the user. The various drugs are smoked, sniffed or injected into the bloodstream. In each case, the chemical is absorbed into the blood and carried to the brain. Short-term risks include having accidents while under the influence of the drug. Overdoses may kill. Brain damage can result from long-term use. But perhaps the most dramatic effect drugs have is the way they can completely take over a person's life. Drug addicts become physically addicted to a drug and become desperate for their next 'fix'. They may commit crimes to obtain money to buy the drug.

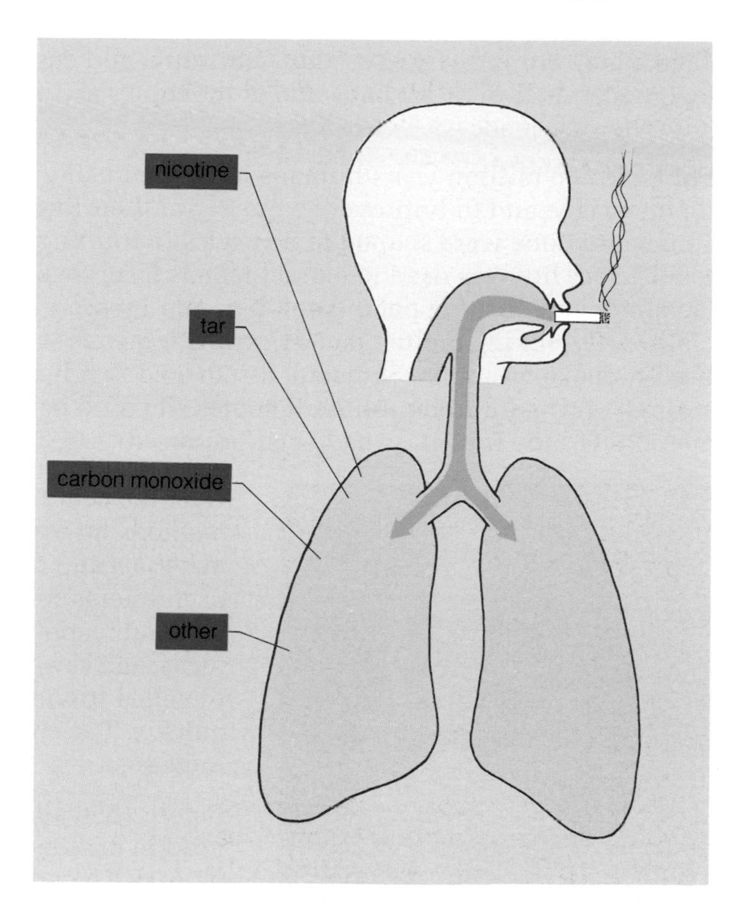

In the future, you may be offered cigarettes, alcohol and possibly even illegal drugs. You will have to make the decision, but you know that:

- all these things contain chemicals to which you may become addicted
- they can all damage your health
- friends who think less of you because you do not want to smoke, drink or take drugs are not real friends.

1. What are the particular dangers involved with: a) injecting drugs, b) drinking alcohol, c) sniffing glue, d) smoking cigarettes? ▲
2. Does the fact that a drug is legal make it safe to use? ▲
3. How can your smoking affect the health of other people?
4. Discuss three reasons why you think people might take drugs when they are not ill. Write them down.

Materials 18

Take a look around you at clothes, furniture, and equipment. Spend a few moments thinking what some of the things are made from, and how they are made.

For over two million years humans have been using tools to help them survive and to improve the quality of their lives. At first, wood, stone, and bone were shaped to be useful in hunting and preparing food. Later, fire was used to extract metals from rock to make jewellery and cooking pots. Wool, hair and fur were collected to make clothes. By mixing molten metals, stronger metals were produced and used to make weapons. Someone discovered that heating sand strongly, turned it to liquid. As it cooled, it could be shaped to make containers - the first glass had been produced.

This bronze incense burner is 3100 years old!

Sir William Perkin.

The first synthetic substance.

Over the centuries, better methods for extracting natural materials and turning them into useful items were discovered. As communications improved, good ideas, and raw materials, travelled round the world more quickly. The speed of progress increased.

Before 1856 all the substances used in making things were natural. They came directly from plants, animals, or the Earth. Then a young chemistry student called William Perkin made a breakthrough which was to be the start of a huge industry. While he was investigating coal tar, he produced a new substance which could be used to dye cloth purple. He called it mauveine, and became rich very quickly by supplying it to the textile industry. Mauveine was the first of many **synthetic** substances to be created. Synthetic means something that is made rather than occurring naturally. Now synthetic materials, such as plastics, are used for many purposes, and their production is a huge international industry.

Waterproof clothing made from Goretex.

For thousands of years heating and mixing have been used to make substances more useful. In this section you will find out some of the background to the physical and chemical processes that are so important to us.

18.1 Solids, liquids, and gases

Starting off

You already know these three words. They are used in everyday language to describe materials. Scientists use the words to describe the **state** of a material. Here are some descriptions of the three states:

We now know that water particles are the same, whether in ice, water or steam. It is the **arrangement** of the particles, and the **forces** between them, which are different in a solid, a liquid and a gas. This table explains the differences:

	SOLID	LIQUID	GAS
	ice	water	steam
ARRANGEMENT of particles	Particles are in a fixed framework	Particles are free to move in the liquid	Particles are free to move anywhere
NUMBER of particles	Many particles in a certain volume	Many particles in a certain volume	Few particles in a certain volume
SPACING of particles	Particles are close together	Particles are close together	Particles are far apart
FORCES between particles	Forces between particles are very strong	Forces between particles are strong	Forces between particles are very weak

1. Use the idea of particles to explain why it is difficult to alter the shape of a solid?
2. Use the idea of particles to explain why it's easier to clap your hands in air than under water? ▲
3. Use the idea of particles to explain why solids and liquids are dense, but gases are less dense? ▲
4. Sort out the speech bubbles to make three descriptions about solids, liquids, and gases.
5. How could you check that the volume of a liquid stays the same, even when its shape changes?

Did you know?

- The average density of the Earth is 5.5 grammes per cubic centimetre.
- The density of the air at the Earth's surface is less than 0.0013 grammes per cubic centimetre.

18.1 Moving particles

Going further 1

Solid

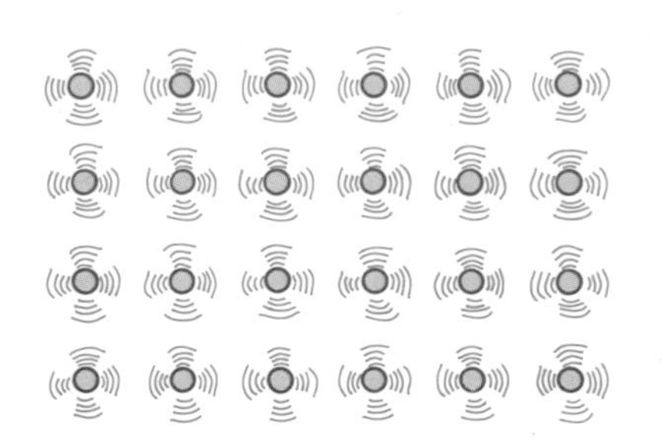

Each particle vibrates around its fixed place in the framework.

Liquid

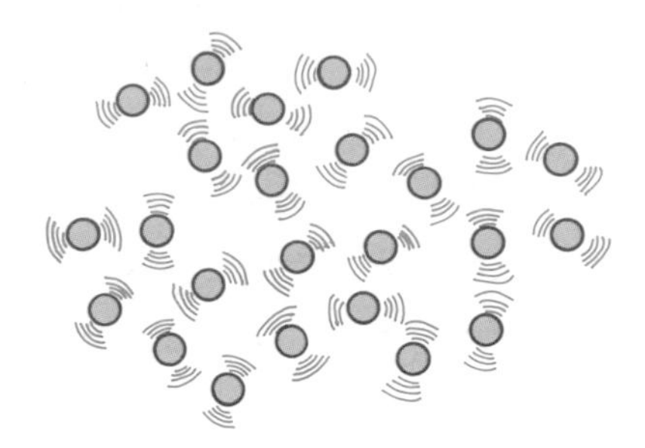

Particles are free to vibrate in any direction.

Gas

Vibrations are large, particles are free to vibrate in any direction.

One of the things this 'particle theory' couldn't explain was how substances behaved when they were heated or cooled. By adding or taking away energy, a substance can be made to change from one state into another. About a hundred years ago a theory explaining this was developed.

The **kinetic theory** suggests that:

- the particles in a substance are always vibrating.
- the size and speed of the vibrations increases with temperature.

Melting If a solid is warmed its particles gain kinetic energy. Their vibrations increase until the vibrating forces overcome the forces holding the particles together in the solid 'framework'. The particles break free and become part of a liquid.

Evaporation If heat energy is supplied to the liquid, the particles in the liquid gain kinetic energy and move further apart from each other. As they do this, the warm liquid becomes less dense and floats towards the surface. In any heated liquid the most energetic particles rise to the top. As their kinetic energy increases, the vibrations will overcome the forces holding the particles together at the surface of the liquid. The particles break through the surface and form a gas above it.

Condensation When heat energy is removed from a gas, its temperature falls. The particles in the gas lose kinetic energy and move closer together. At a certain stage, they will come close enough to form a liquid.

Freezing As a liquid cools down its particles lose kinetic energy. As the vibrations of the particles get smaller, at a certain temperature they are overcome by the forces attracting particles to each other. They start to move into the 'framework' formation which is typical of a solid.

Uranium atoms arranged in a perfect hexagonal shape around a central atom in a crystal. Although they are locked in this framework, they are always vibrating.

Did you know?

- 'Kinetic' means 'moving'. The first cinemas were called 'kinemas' – places where moving pictures were shown.

1. Describe what happens to the particles of a liquid as it is heated. ▲
2. What is the name for the change from gas to liquid? Give an example of where you have seen this happening. ▲
3. Where does the heat energy come from to melt an ice cube which is left on a kitchen work surface?
4. **Try to find out how** the kinetic theory explains
   a) diffusion (for example, the spread of a smell)
   b) how a solid dissolves in a liquid.

18.1 Putting a theory to work

You can use the kinetic theory to explain some of the things that you have already observed about materials.

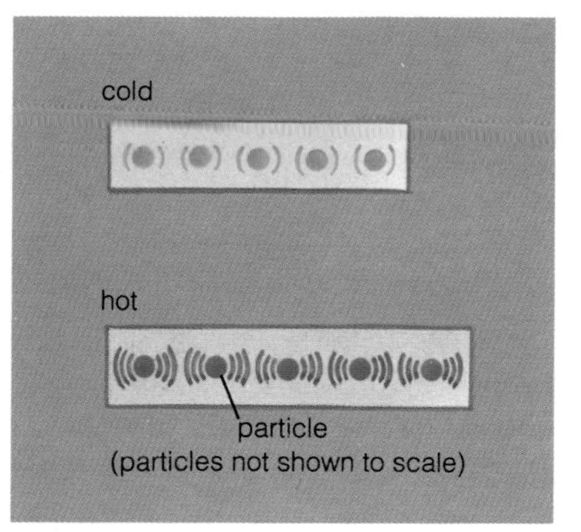

Expansion Particles in the cold bar vibrate, but when the bar is heated they vibrate much more. Because the vibrations are bigger, each particle takes up more space. The particles have to move away from each other to make space for the bigger vibrations. All these small movements in the millions of particles in the bar add up to make a difference in the size of the bar. It gets longer and wider as the temperature increases.

Different substances expand a different amount for each °C rise in temperature. Liquids and gases also expand when heated.

Convection When the beaker is heated by a burner, the water nearest to the flame warms up first. Its particles gain kinetic energy. They vibrate more and so move further apart. The warmed water will be less dense than the cool water above it, and will float up through it. A flow of warm water from the base of the beaker is started. This is called a **convection current.**

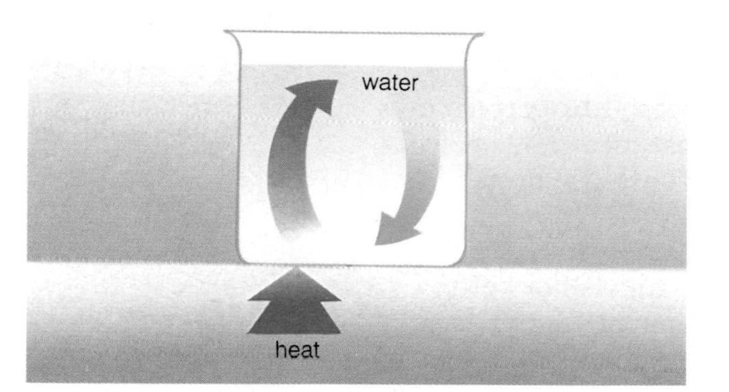

Conduction The bar starts off cold. When a burner is placed under the left-hand end, the particles there gain kinetic energy. They vibrate more.

The increased vibrations of these particles make them 'bump into' particles to the right and cause them to vibrate more. The same thing happens along the bar. As each particle is hit by the particle on the left, its own vibrations increase. The 'message' is passed along the bar. But remember, the kinetic theory says that bigger vibrations mean an increase in temperature. The end of the bar towards the right must now be hotter than when it started. Heat has been conducted along the bar.

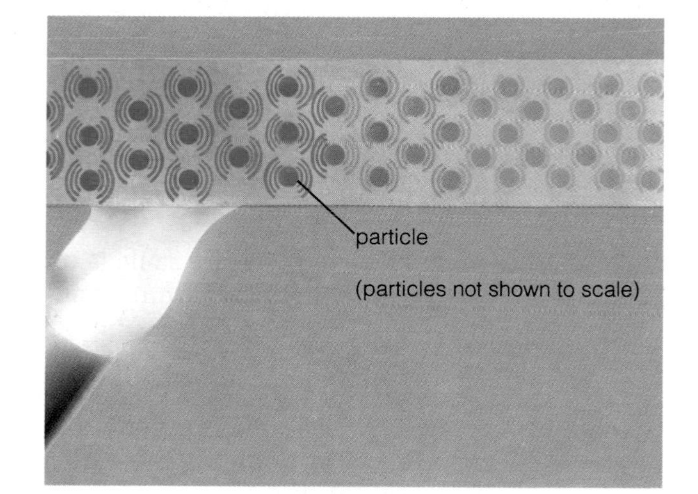

1 Describe why a gas expands when it is heated. ▲
2 When a 30 cm aluminium bar is heated by 50 °C, its length increases by 0.5 mm. If a 60 cm bar of the same diameter was heated by the same amount, by how much would its length increase?
3 Explain how heat is conducted through materials. ▲
4 Cooking pans sometimes have bases made from copper (good heat conductor), or steel (poor heat conductor). What effect might these different materials have on the way food cooks?
5 Explain why warm water is less dense than cold water.

Did you know?

- In a microwave oven, water molecules are given energy by the microwaves. As they vibrate more, they cause the other food molecules to vibrate. The temperature rises and the food is quickly cooked.

18.1 Pressure, volume, and temperature

An odd prediction

The kinetic theory suggests that as a particle increases in temperature the size of its vibrations get larger. The opposite is also true. As a particle is cooled down the size of its vibrations gets smaller.

If we keep on cooling the particle its vibrations should keep on getting smaller until, at a certain temperature, the particle stops vibrating completely.

Now, you can't have a vibration which is smaller than nothing. So the kinetic theory predicts that this should be the lowest temperature which it is possible to reach. This is called **absolute zero**.

By studying the way the volume of a gas decreased as it was cooled down, scientists predicted that absolute zero would be at about minus 273°C.

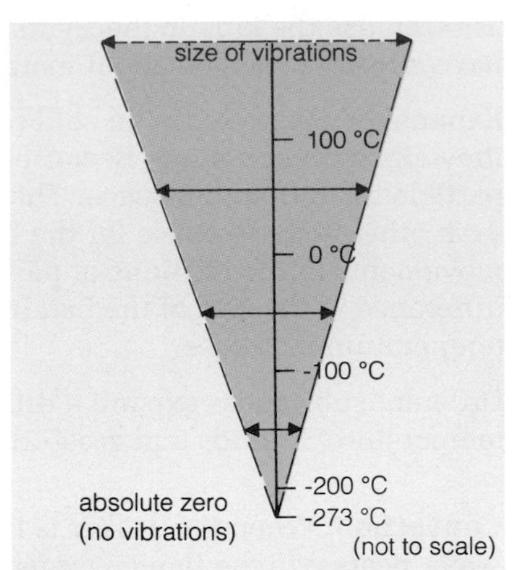

Absolute temperatures are measured on scale starting at minus 273°C. This temperature scale is named after Lord Kelvin who first produced it. A change in temperature of one kelvin (K) is the same as a change of 1°C.

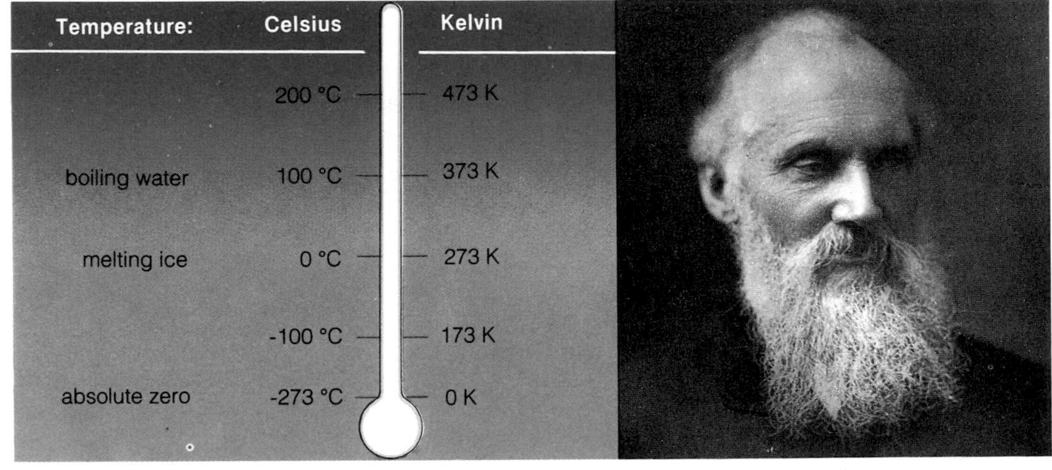

Lord Kelvin (1824-1907).

The absolute (or kelvin) temperature scale is useful for looking at the relationship between the pressure, volume and temperature of a gas.

For instance, when the pressure of a gas is kept constant, if the absolute temperature doubles, the gas will double in volume.

Sometimes, a gas might be sealed in a container so that its volume cannot change. In this case, doubling the absolute temperature would double the *pressure* of the gas.

Did you know?

- An aerosol can is sealed. If it is heated, the pressure of the gas inside will increase until the can explodes.

1. What would happen to the particles in a gas if you cooled them to absolute zero? (Note: the size of the actual particles stays the same.)
2. Frozen carbon dioxide (dry ice) changes directly from a solid to a gas. Use the kinetic theory to describe what would happen if a little dry ice was put into a sealed plastic bag and left to warm up.
3. If the absolute temperature of the gas in a balloon is halved, by how much will its volume change? ▲
4. Why does a bicycle pump get warm as you pump up the tyres?

18.2 Atoms inside and out

Starting off

You will already be familiar with the idea of atoms, elements, mixtures, and compounds. Here is a reminder of what these words mean in science:

Atoms are tiny particles which are the building blocks for all substances, whether they are solids, liquids or gases.

Elements are the simplest substances. They contain just one type of atom. The atoms of one element are different from the atoms of another element. There are about 100 elements, including, for example, oxygen, sodium and mercury.

Mixtures are formed when the particles of different substances mingle, but do not combine to make new chemicals.

Compounds are formed when two or more elements combine to make a new chemical.

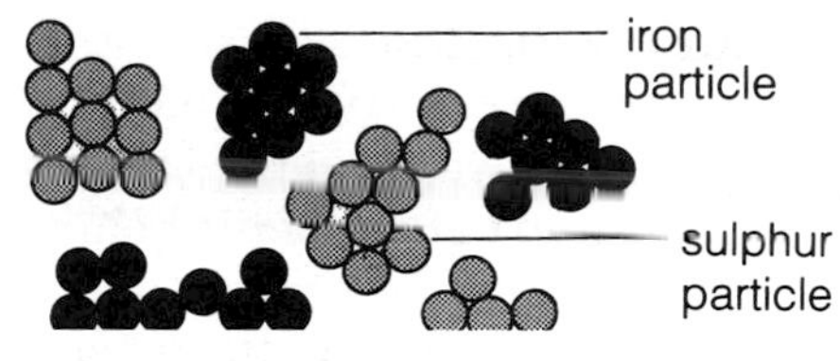

A mixture.

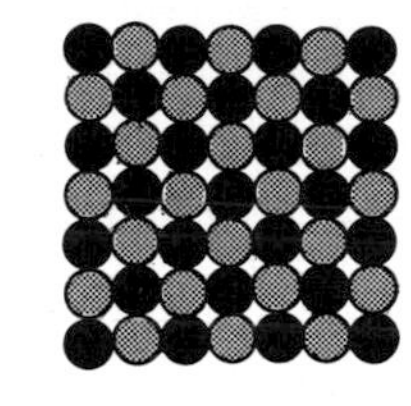

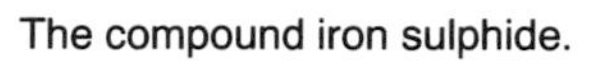

The compound iron sulphide.

What makes different elements behave differently?

For hundreds of years atoms were thought to be the smallest particles. Then it was discovered that atoms themselves are made up from even smaller particles called **protons**, **neutrons**, and **electrons**. The atoms of each element have certain numbers of protons, neutrons, and electrons. This makes the atoms of different elements behave differently.

An oxygen atom has
8 protons, 8 neutrons, and 8 electrons.

A sodium atom has
11 protons, 12 neutrons, and 11 electrons.

A mercury atom has
80 protons, 121 neutrons, and 80 electrons.

Protons and neutrons are more massive than electrons.

The protons and neutrons are in a group at the centre of the atom, known as the **nucleus**. Electrons circle around the nucleus very quickly. The nucleus is tiny compared to the rest of the atom, smaller than a grain of sugar in the middle of an ice rink.

electron (– charge)
proton (+ charge)
neutron (no charge)

Depending on the arrangement of electrons around a nucleus, atoms of different elements may tend to 'lose' or 'gain' electrons. If the tendency is strong, then the element is **reactive**. If it is weak, the element is **unreactive**.

1 Explain the difference between a mixture and a compound. ▲
2 Name six different elements, and for each, add what you know about it.
3 Copy and complete this sentence:
At the centre of the atom is the _________. It contains _______ and __________. ____________ make up the rest of the atom. ▲
4 From the examples above, what do you notice about the numbers of protons and electrons in an atom? ▲

Did you know?

- The atoms of a particular element always have the same number of protons, but sometimes they have a different number of neutrons. These alternative versions of an element are called **isotopes**.

18.2 The periodic table

Going further

Around 1870, a Russian scientist called Dmitry Mendeleev was attempting to classify the 63 different elements that were known at that time. He was looking for a pattern which would help explain why certain groups of elements behave in a similar way. He tried listing the elements in the order of what he called their 'atomic weight'. He found that elements with similar properties were **periodic**. They appeared on his list in a regular pattern.

Mendeleev designed a useful way of presenting the elements in a table. He called it the 'Periodic Table of Elements'. Our periodic table today is similar, but puts the elements in order of the number of protons in an atom. It includes the many elements discovered since 1870.

Each element has its own symbol which is a shortened version of its name in English (or sometimes in another language).

Beryllium x
Boron o
Carbon Δ
Nitrogen
Oxygen
Fluorine
Sodium
Magnesium x
Aluminium o
Silicon Δ

The symbols show which elements have similar properties.

Number of protons

	Group I	II											Group III	IV	V	VI	VII	O
Non-metals			1 H hydrogen															2 He helium
Metals	3 Li lithium	4 Be beryllium											5 B boron	6 C carbon	7 N nitrogen	8 O oxygen	9 F fluorine	10 Ne neon
	11 Na sodium	12 Mg magnesium	← transition elements →										13 Al aluminium	14 Si silicon	15 P phosphorus	16 S sulphur	17 Cl chlorine	18 Ar argon
	19 K potassium	20 Ca calcium	21 Sc scandium	22 Ti titanium	23 V vanadium	24 Cr chromium	25 Mn manganese	26 Fe iron	27 Co cobalt	28 Ni nickel	29 Cu copper	30 Zn zinc	31 Ga gallium	32 Ge germanium	33 As arsenic	34 Se selenium	35 Br bromine	36 Kr krypton
	37 Rb rubidium	38 Sr strontium	39 Y yttrium	40 Zr zirconium	41 Nb niobium	42 Mo molybdenum	43 Tc technetium	44 Ru ruthenium	45 Rh rhodium	46 Pd palladium	47 Ag silver	48 Cd cadmium	49 In indium	50 Sn tin	51 Sb antimony	52 Te tellurium	53 I iodine	54 Xe xenon
	55 Cs caesium	56 Ba barium	57 La lanthanum	72 Hf hafnium	73 Ta tantalum	74 W tungsten	75 Re rhenium	76 Os osmium	77 Ir iridium	78 Pt platinum	79 Au gold	80 Hg mercury	81 Tl thallium	82 Pb lead	83 Bi bismuth	84 Po polonium	85 At astatine	86 Rn radon
	87 Fr francium	88 Ra radium	89 Ac actinium	28 other elements including uranium and plutonium														

Group I contains the **alkali metals.** They

- are soft enough to be cut with a knife
- react violently with water
- corrode quickly in air, and have to be stored in oil.

Group VII contains the **halogens**. They

- react easily with metals
- are all coloured
- are all poisonous and corrosive
- dissolve in water and can be used as bleaches.

Group O contains the **noble gases**. They

- are all very unreactive
- glow when electricity is passed through them and can be used in lighting.

1 How are the elements ordered in the modern periodic table? ▲
2 Name as many of the alkali metals, halogens and noble gases as you can, and list them with their symbols. ▲
3 How many protons are there in an atom of chlorine, gold and mercury? ▲
4 List three of the radioactive elements.
5 Which of the noble gases do you know is used in lighting?

Did you know?

- The test of Mendeleev's table came in 1871. In order to make the groupings, he had to leave a gap in table between gallium and arsenic. He predicted that there was an unknown element which would fill the gap. In 1886 this element was discovered, and called germanium.

18.2 Combining elements

For the enthusiast

You know that a compound is made up from the atoms of two or more elements. The **formula** of a compound is written using the symbols of the elements it contains.

Water molecules contain 2 hydrogen atoms, and 1 oxygen atom. Water has the formula H_2O. Salt contains 1 sodium atom and 1 chlorine atom and has the formula NaCl.

How compounds are formed

An atom's electrons are the key to how it joins, or **bonds**, with another atom.

In some bonds, electrons are shared between atoms. These are called **covalent bonds**. A group of atoms joined with covalent bonds is called a molecule.

Examples of molecules are:

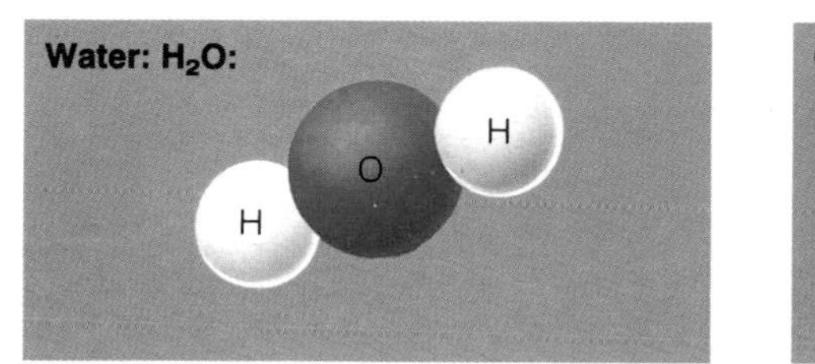

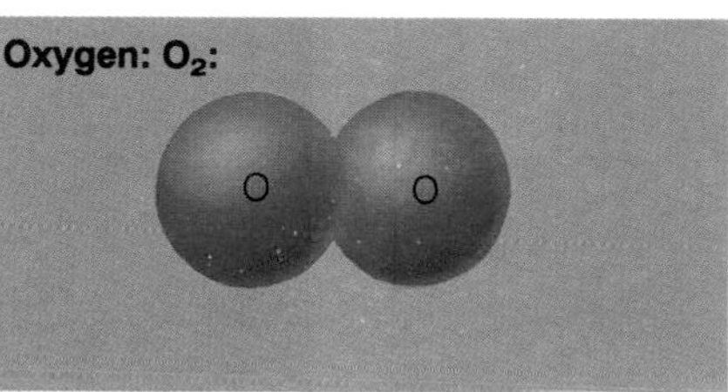

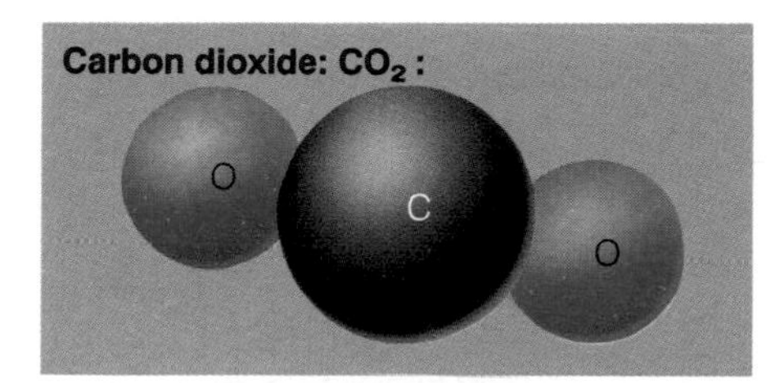

This electron-sharing bond only happens between non-metals.

In some bonds one or more electrons may be switched between atoms. When iron and sulphur combine, for example, iron atoms each lose two electrons and sulphur atoms gain two electrons. This is called an **ionic bond**.

An atom with electrons added or taken away is called an **ion**. So the compound iron sulphide contains sulphide ions and iron ions! This transfer of electrons between atoms of different elements usually happens between a metal and a non-metal.

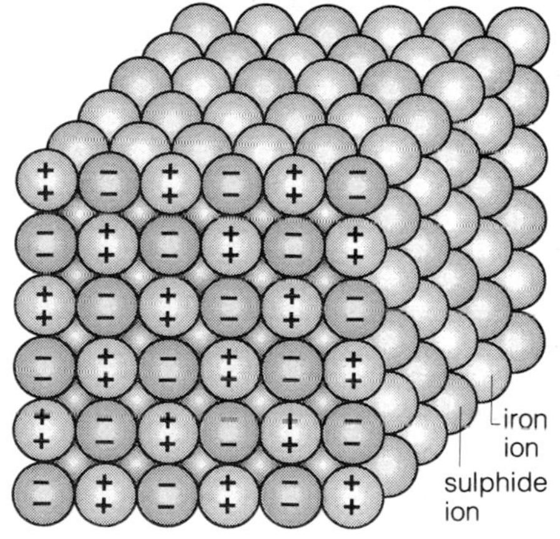

Iron sulphide.

Did you know?

...how the compounds get their names

- **Compounds ending in '-ide'** When a metal combines with a single non-metal, the metal keeps its name, and the non-metal changes its ending to '-ide'. For example, oxygen becomes **oxide**, fluorine becomes **fluoride**, sulphur becomes **sulphide**, and bromine becomes **bromide**.
- **Compounds ending in '-ate'** always include a group consisting of a non-metal atom linked with oxygen atoms. Here the non-metal changes its ending to '-ate'. For example:
  the **sulphate** group (SO_4) has 1 sulphur linked with 4 oxygen atoms
  the **nitrate** group (NO_3) has 1 nitrogen linked with 3 oxygen atoms.

1 Describe the two sorts of bonds that can be formed between elements. ▲
2 What is the chemical name, and the formula for salt? ▲
3 What elements make up the following compounds? ▲
a) iron sulphide b) iron sulphate
4 What are the names of these zinc compounds?
a) $ZnCO_3$ b) ZnO
c) $ZnSO_4$ d) ZnS

18.3 Changes

When an ice cube melts, it is easy to make the water freeze again. When a spoon of salt dissolves in water, you can dry out the solution to retrieve the salt. Melting and dissolving are examples of **physical changes**. No new chemical substances are produced in this type of change.

When iron rusts, or when toast burns, a new chemical compound is made. It is difficult to reverse the process (unfortunately!). These are examples of **chemical changes**, also known as chemical **reactions**. If you're not sure whether a chemical reaction has taken place, use this checklist.

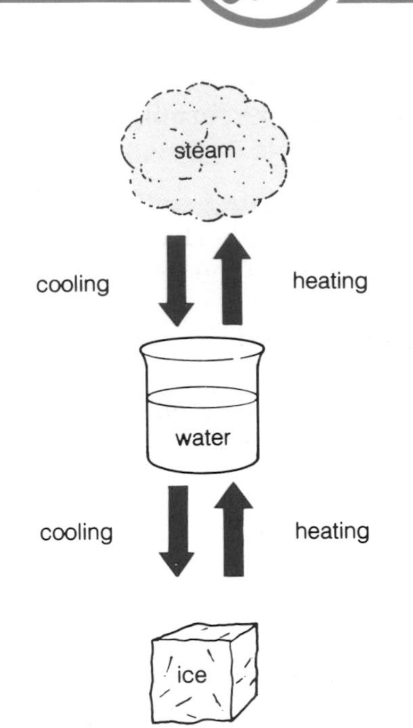

Physical change.

In a chemical reaction:

- a new substance is made which usually looks different from the substances you started with

and

- energy is taken in or given out (for example, heat energy)

and

- the change is hard to reverse.

When iron and sulphur are mixed and heated together, a new substance is formed called iron sulphide. This can be written in a **word equation**:

iron + sulphur $\xrightarrow{\text{heat}}$ iron sulphide

Or it can be written in symbols in a **chemical equation**:

$Fe + S \rightarrow FeS$

The substances at the start of a reaction are called the **reactants**.

The substances left at the end of reaction are called the **products**.

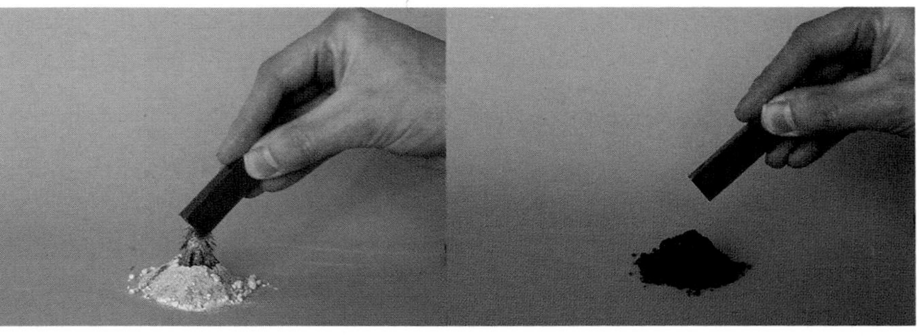

In a *mixture* of iron and sulphur, the iron can be removed.

In a compound, iron sulphide, separation is much more difficult.

Although you have to heat the iron and sulphur to start up this reaction, it gives out heat once it gets going. A reaction which gives out heat is called **exothermic**. Exothermic reactions can be very dramatic! Fireworks are designed by scientists who choose chemical substances and mix them in just the right amounts to produce light and sound energy, as well as heat energy.

Some reactions take in heat while they are happening. These are called **endothermic** reactions.

Did you know?

- Cooking involves many physical and chemical changes. When you cook an egg, the white changes very quickly to become more solid and opaque white. This is a chemical change and cannot be reversed.

1. Describe the difference between a chemical change and a physical change. ▲
2. What is the name given to the substances at the start of a reaction? ▲
3. Write down the three things that happen in chemical reactions. ▲
4. Do you think that reactions are important in our lives? Why do you think this? *Hint:* Think about the things around you and what they have been made from. Think about the processes happening in your body.

18.3 Useful reactions

Reactions are happening all the time in nature. They are essential for life. Respiration, photosynthesis, and fermentation are all examples of chemical reactions.

Fermentation

Yeast grows on the skins of grapes. If the skins are broken, the yeast feeds on the glucose in grape juice, and turns it into an alcohol (called ethanol), and carbon dioxide. This process is known as fermentation and is used in wine-making.

In bread-making, yeast is used to make the bubbles in dough. As the mixture starts to ferment, carbon dioxide is produced which gets trapped in the dough and causes it to rise. During baking the yeast is killed and the alcohol is evaporated.

Like all reactions, fermentation can be shown in a word equation:

$$\text{glucose} \xrightarrow{\text{yeast}} \text{alcohol} + \text{carbon dioxide}$$

Other reactions can be set up to make the things that we need.

Ammonia production

Ammonia is a very useful chemical substance. It is used in cleaning fluids around the house, and in making some fertilizers and plastics. It is produced by mixing the gases hydrogen and nitrogen under pressure. Hot iron is needed also, to make the reaction happen.

The word equation for the reaction is:

$$\text{nitrogen} + \text{hydrogen} \xrightarrow{\text{hot iron}} \text{ammonia}$$

Hot iron is not a reactant or a product in the reaction, but needs to be there to make it happen. Substances which help a reaction along like this are called **catalysts**. At the end of a reaction, a catalyst is chemically the same as at the start.

Some catalysts used in industry.

1 Explain how the bubbles in bread are formed. ▲
2 Give four examples of reactions in nature, and write down the word equation for each.
3 Describe in your own words what a catalyst does. ▲
4 Why might it be dangerous to leave crushed grapes in a sealed container? ▲
5 Give another example of an enzyme. What reaction does it speed up?

Did you know?

- In living things like yeast, the catalyst is called an **enzyme. Zymase** is the name of the enzyme in yeast which works in fermentation.

18.3 Reacting with oxygen

Elements combine with oxygen in a process called **oxidation** to form **oxides**. Here are some examples of how oxides are formed:

Burning

Burning, also called **combustion**, is actually a type of oxidation reaction. During burning there is a flame, a glow, or a bright flash, and heat energy is given out. The main product of the reaction is an oxide. Air contains only about a fifth oxygen, and substances can be made to burn more quickly by replacing air with pure oxygen.

Many oxides are invisible gases. When carbon burns, carbon dioxide gas is produced. The word equation is:

carbon + oxygen → carbon dioxide

Oil and coal are used as fuels because they produce a lot of heat energy when they burn. Other substances give out bright light too. Magnesium burns with a white flash, and calcium with a red flash which makes them useful to people who make fireworks.

Corrosion

Metals may react with air to make oxides without burning. This process is called **corrosion**. When iron corrodes, rust is produced. The word equation for this reaction is:

iron + oxygen → iron oxide

When metal atoms react with oxygen atoms to form oxides, they lose electrons to the oxygen atoms. Remember that reactive metals lose electrons more easily, and so will be more likely to form oxides. The alkali metals are so reactive that they have to be stored in oil. They corrode very quickly in air.

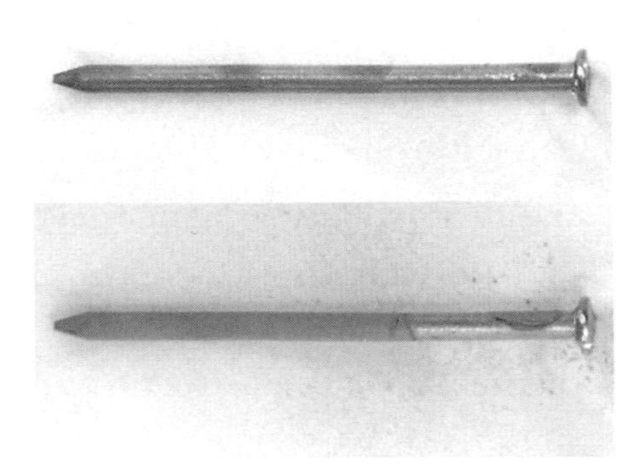

More iron reacts to form iron oxide as time goes on.

Did you know?

- Corrosion is responsible for damage costing millions of pounds every year. Mixtures of metals, called **alloys**, have been developed which are corrosion-resistant. Stainless steel is an example of a strong, corrosion-resistant alloy.
- Plastics have replaced metal for some purposes because they don't corrode. Drainpipes are an example.

1 What is the name of the process in which elements are combined with oxygen? ▲
2 Write a word equation for burning sulphur in oxygen to make sulphur dioxide.
3 Can you name an oxidation reaction that occurs in all living organisms?
4 Plastic bicycles have been developed, but have not become popular. Can you think of some of their advantages and disadvantages?

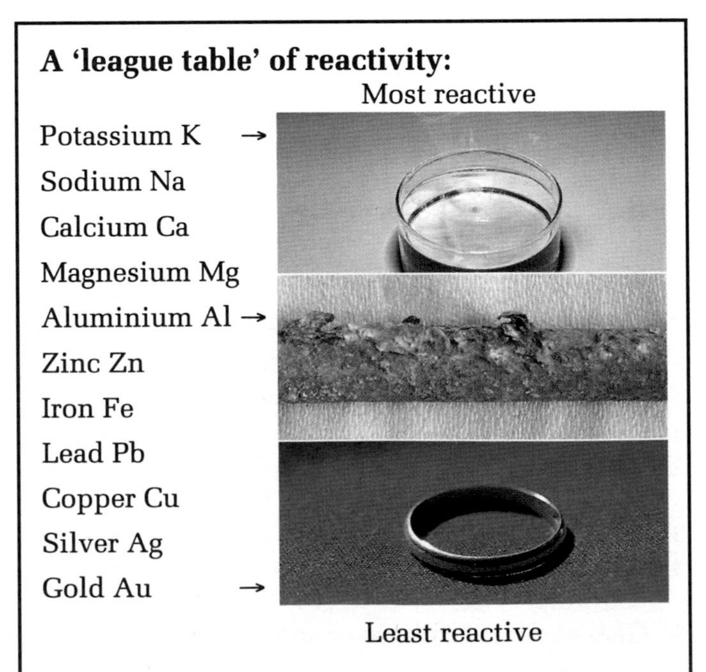

A 'league table' of reactivity:

Most reactive

Potassium K →
Sodium Na
Calcium Ca
Magnesium Mg
Aluminium Al →
Zinc Zn
Iron Fe
Lead Pb
Copper Cu
Silver Ag
Gold Au →

Least reactive

18.3 Reactions: fast or slow?

Going further 3

Whether a reaction happens naturally or is set up, it is important for us to know how to control it. We can make a reaction happen faster or more slowly by changing different things about it. A fast **rate** of reaction means that more reactants are turned into products in a certain period of time. If there is a big demand for the product, then this is obviously a good thing. However, some reactions give out a lot of energy and it may be necessary to slow them down so that the violence of the reaction does not cause damage.

You have already looked at catalysts, which speed up reactions. Various other factors can affect the speed of a reaction.

Things which affect reaction rates can be found by altering single factors in reactions and comparing the results. In other words, you can change one variable, and measure its effect on reaction rate.

If marble chips are added to dilute hydrochloric acid, they fizz and dissolve. The word equation for this reaction is:

calcium carbonate + hydrochloric acid → calcium chloride + water + carbon dioxide

Carbon dioxide gas makes the fizz, and so it is fairly easy to see any difference in the rate at which it is made.

Temperature

Increasing temperature speeds up the rate of reaction.

Concentration

If the concentration of the hydrochloric acid is increased, then the reaction rate speeds up.

Surface area

If the chips are broken into smaller lumps, then a greater surface area will be in contact with the acid.

What effect do you think surface area has on reaction rate?

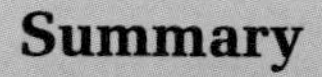

Summary

Particles must hit each other for a reaction between them to take place:

Reaction rates are increased when something happens which

- makes the different reactant particles collide with each other more often. This could be by:
  - increasing the surface area of a solid
  - increasing the concentration of a solution
  - increasing the pressure of a gas.
- gives the particles more energy so that the collisions are more likely to be successful. This could be by:
  - increasing the temperature.

Catalysts make it more likely that a collision is successful without needing more energy.

1. Why do think that it is useful to know how to control reaction rate? ▲
2. How could you increase the surface area of the marble as much as possible?
3. List four factors which affect the rate of a reaction. ▲
4. Explain how particles are affected by these rate-changing factors.▲
5. **Try to find out** the name of some catalysts, including enzymes, and the reactions in which they are used.

18.3 Radioactivity

The elements that you have looked at so far stay the same throughout the reaction, even though they are combined in completely different ways. However, some elements *do* change into different elements. Uranium, plutonium, radon and radium are examples. These elements are **radioactive**, and their atoms naturally break up (**decay**) to produce elements with smaller atoms. When this happens, they give out, or emit, **radiation**.

There are three sorts of radioactive emission:

alpha particle (α)	beta particle (β)	gamma ray (γ)
2 protons and 2 neutrons	an electron	a wave similar to an X-ray

- alpha particles (α) consist of two protons and two neutrons. They soon slow down if something gets in their way. They are stopped by a few centimetres of air, and they cannot pass through skin.
- beta particles (β) are fast-moving electrons. They can pass through skin. They can be stopped by an aluminium sheet a few millimetres thick.
- gamma rays (γ) are high energy rays rather like x-rays or microwaves. They can pass through most thin objects. It takes a few centimetres of lead to stop them.

Radiation can be detected using a **Geiger counter**.

Some radioactive substances occur naturally in rocks and some radiation reaches us from space. There is always some radiation around. This is called **background radiation**. It is normally at a safe level, though it is greater in some areas than others.

You will have heard about radioactivity and some of its dangers and uses.

- Nuclear power stations use the heat from radioactive reactions to generate electricity.
- Radiation has many uses in medicine:
  - Gamma rays can be directed at cancer cells to kill them.
  - Weak radioactive substances can be fed to a patient in food, and the path of food traced through the digestive system. It is possible to detect any blockages or other problems in this way.
- Radioactive 'tracer' substances are also used in industry to find leaks in pipes, for example.

Radiation can damage living things because it causes atoms to lose or gain electrons. They turn into ions, and the process is called **ionization**. Ionization can upset the balance of living processes and cause damage to cells and tissues.

Radiation levels need to be monitored all the time, especially around nuclear power stations or places where radioactive materials are stored.

Did you know?

- Some radioactive elements are found naturally, but others are made in a laboratory.

1. How are radioactive elements different from other elements? ▲
2. What instrument is used to detect radiation? Where do you think it may be used? ▲
3. How can radiation harm living things? ▲
4. Name three jobs in which people will need to wear clothing to protect them from radiation?
5. Name and describe the three different types of radiation. ▲

18.4 Where materials come from

When our ancestors first made things, they used the materials they found around them such as rocks, wood, bone, and animal skins. As time went on, they learned how to alter these **natural** materials. They learned how to make new materials.

But even now, all these synthetic materials are made from substances which come from the Earth's crust, from the sea or from the atmosphere.

These three charts show the proportions of elements by their weight.

Elements in the Earth's crust

The Earth has a diameter of about 12 750 km. But we can only get at the materials in about the top 5 km. Even this, though, contains a vast treasure chest of materials.

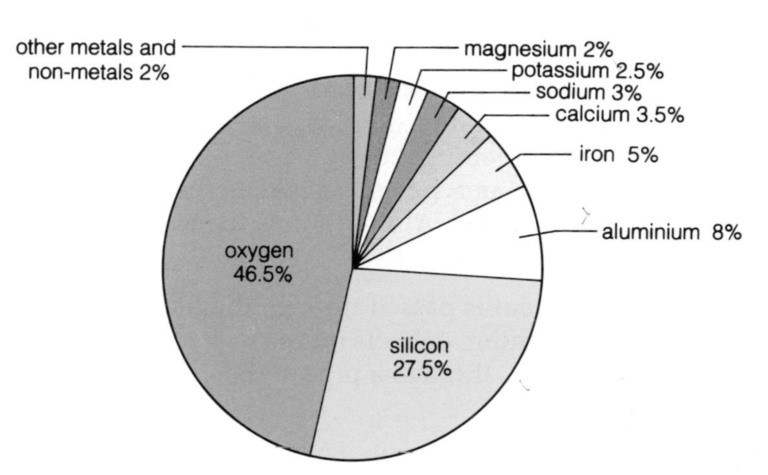

Elements in the sea

As you would expect, most of the sea (nearly 97%) is water. The amount, or abundance, of these elements may vary. Because heat causes high evaporation of water from the Red Sea, it has a slightly higher salt concentration.

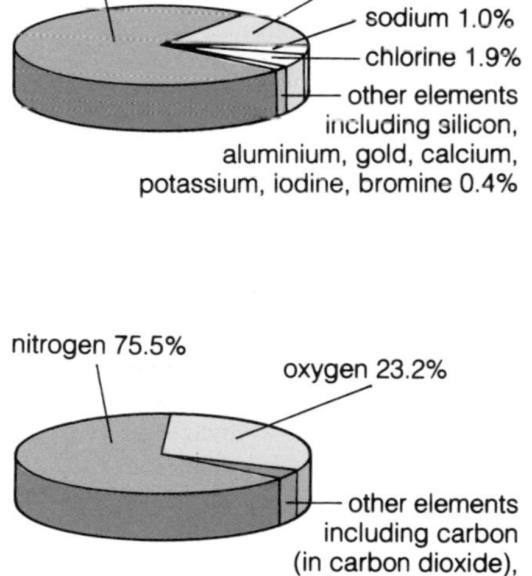

Elements in the air

This is what you breathe in! Which of the gases does your body use?

Electrolysis is often used to separate metals from the compounds in which they are naturally found. The impure metal, called an **ore**, is melted or dissolved in a liquid so that the ions in the compound are free to move. Then electrodes are dipped into the liquid. The positively charged metal ions collect on the negative electrode, or **cathode**. This process uses up a lot of electricity and so is expensive.

1 Make a list of elements in the Earth's crust in order of quantity. Put the most abundant at the top. ▲
2 Which is a) the most **abundant** element on Earth
b) the second most **abundant** element on Earth? ▲
3 Why couldn't we all get rich by extracting gold from the sea?
4 If you were told that sea water contained significant amounts of chlorine, bromine and fluorine, what two other elements might you also expect to find there?

Did you know?

- Each cubic kilometre of sea water contains 6 tonnes of gold (worth about £40 000 000).

18.4 Packaging 1: Aluminium cans

Going further 1

Young people buy billions of soft drinks each year. Drinks would be much cheaper if you could buy only the drink. But of course, you need the container too! The manufacture of containers in which to sell drinks is a multimillion pound industry.

Aluminium cans, 'tin' cans, glass and plastic bottles, and 'paper' cartons are all popular methods of packaging.

Making aluminium

Aluminium occurs in the Earth's surface as the **ore**, bauxite. Bauxite is aluminium oxide (Al_2O_3) and has a very high melting point of 2045°C. The bauxite is first dissolved in a solvent to reduce the melting point to about 950°C. Aluminium is then obtained by **electrolysis**.

When the electric current is passed through the hot solution, molten aluminium collects on the **cathode**. It is 'tapped off' through a pipe at the bottom of the cell.

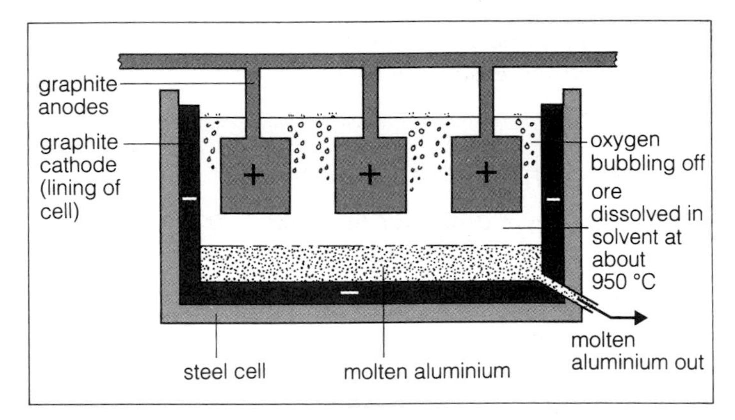

Electrolytic cell for making aluminium.

Making the cans

The process for making the can is called 'drawing and wall ironing' (**DWI** for short).

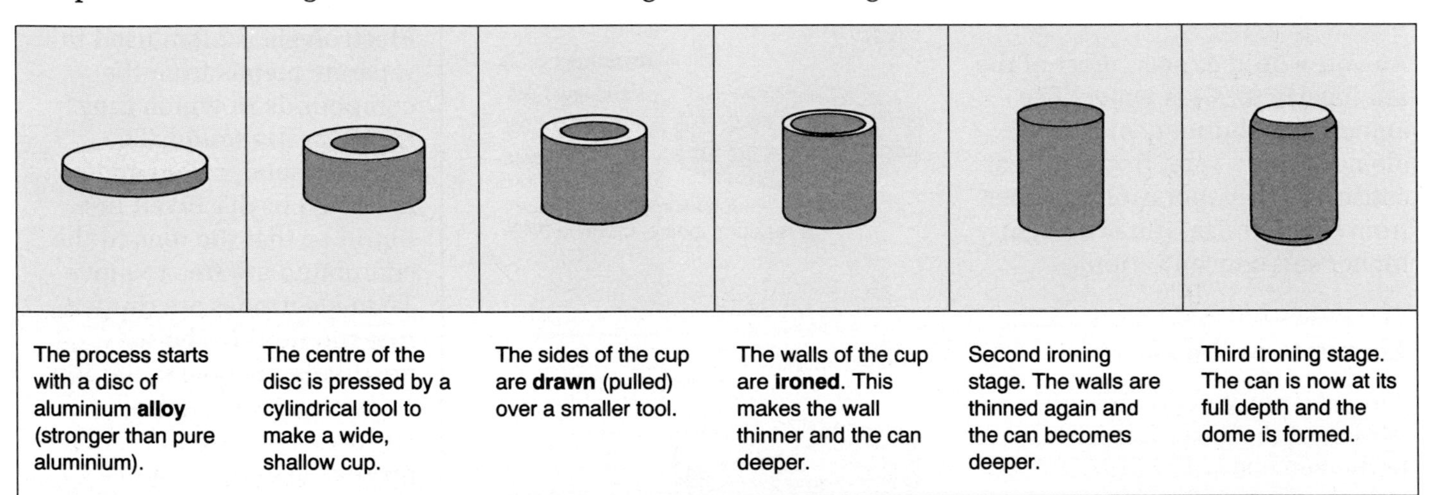

The top of the can is then trimmed off. Tops for the cans are made in a separate process. The can maker then sends the can and top to the drinks manufacturer for filling and sealing. This is all done by machines.

1. In the electrolysis of aluminium oxide, aluminium collects at the cathode of the cell. Which element collects at the anode? ▲
2. The electrolytic cell is made of steel (which melts at about 1500°C). What might be a problem in using it to hold molten bauxite? How is this overcome? ▲
3. What happens to the volume of aluminium alloy during the six stages of the DWI process from disc to can? ▲

Did you know?

- Some can-making machines make over 2000 cans per minute.

18.4 Packaging 2: Steel cans

Going further 2

'Tin' cans are actually made from **steel**, processed from **iron**. They always used to have a coating of **tin** to prevent corrosion. Other coatings are often used now.

Making iron Iron is rarely found in the Earth's crust as the pure element. It usually occurs as a rocky **ore**. The iron ore is converted to iron in a **blast furnace**. One of the ores used contains an iron oxide called **haematite** (Fe_2O_3).

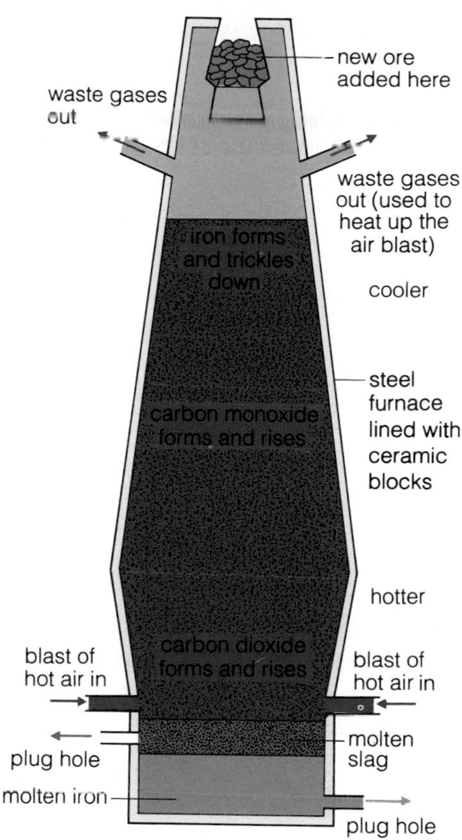

Blast furnace

The blast furnace

Haematite is fed into the top of the furnace along with limestone (calcium carbonate, $CaCO_3$) and coke (carbon). Hot air is 'blasted' through the base of the furnace.

A number of chemical reactions occur during the process.

1 The coke burns in the oxygen of the air to form carbon dioxide.

carbon + oxygen → carbon dioxide

2 The carbon dioxide (CO_2) reacts with more carbon to form carbon monoxide (CO).

3 The carbon monoxide reacts with the haematite to form iron and carbon dioxide.

carbon monoxide + iron oxide → iron + carbon dioxide

At this stage the iron still contains lots of impurities, mainly carbon and silica (silicon oxide, SiO_2). This is where the limestone is needed.

4 The heat of the furnace changes the limestone into calcium oxide and carbon dioxide.

5 The calcium oxide reacts with the silicon oxide impurities in the iron, to produce calcium silicate.

The molten calcium silicate, called **slag**, is skimmed off the top of the molten iron. The iron (containing about 4% carbon) is run off from the bottom into moulds to form ingots of cast iron. Cast iron is very brittle and so it is converted to steel.

Different types of steel are made by processing the iron to reduce the amount of carbon to around 1 % and by adding small amounts of other elements. Scrap steel that is being recycled is added to the mixture.

Making cans Traditional 'three piece' steel cans are made from a rectangle of tinned sheet steel. This is rolled into a cylinder and welded. The bottom disc is then welded into place. The can is then sent to be filled and sealed. Modern 'two piece' steel cans are made by a drawing and redrawing process. This is similar to the first three stages in aluminium can production.

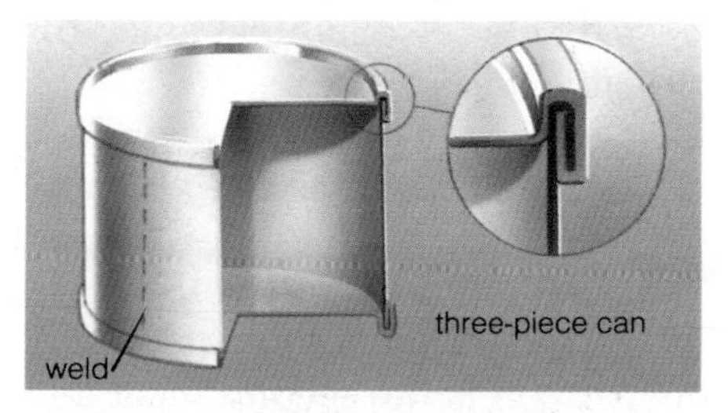

Three-piece steel can.

1 Stages 1 and 3 above gave examples of 'word equations' to describe the chemical reactions. Write word equations for all 5 stages. ▲
2 The seams of three piece steel cans used to be joined with lead. What might the reasons be for using welded joints now?
3 The outside of a blast furnace is itself made of steel. How is it prevented from melting with the heat of the molten iron? ▲
4 What is iron ore? ▲

Did you know?

- New 'tin free' steels have a coating of chromium and chromium dioxide.

18.4 Packaging 3: Bottles

For the enthusiast

Bottles are also widely used as containers for drinks.

The first bottles were made of glass. Plastic bottles became popular in the 1970s. Early plastic bottles, though, were not suitable for fizzy drinks. The carbon dioxide gas molecules escaped through the wall of the bottle!

Bottle-making machine

Glass bottles

Glass is a compound called sodium silicate (Na_2SiO_3). To make glass, silicon oxide (from sand) and sodium carbonate, are heated in a furnace to 1500°C. Scrap glass that is being recycled is added to the mixture. The silicon oxide and sodium carbonate react to form sodium silicate.

silicon oxide + sodium carbonate → sodium silicate

The red hot molten glass oozes through pipes in the bottom of the furnace tank. It is chopped into short lengths and dropped into the moulding machine below the furnace. Compressed air is then blown into the moulds to blow the glass into the shape of the mould.

Plastic bottles

Plastic materials all use oil as their raw material.

Ethene, whose molecules consist of 2 carbon and 4 hydrogen atoms can be obtained from oil. Under high pressure and temperature, ethene molecules join up to make a chain. The new molecule is called **polyethene** ('poly' means many), or polythene. This consists of long chains of carbon and hydrogen atoms. The long chain molecules are typical of plastics, and are called **polymers**.

The latest plastic bottles used for containing fizzy drinks are made from **p**oly**e**thene **t**erephthalate (**PET** for short).

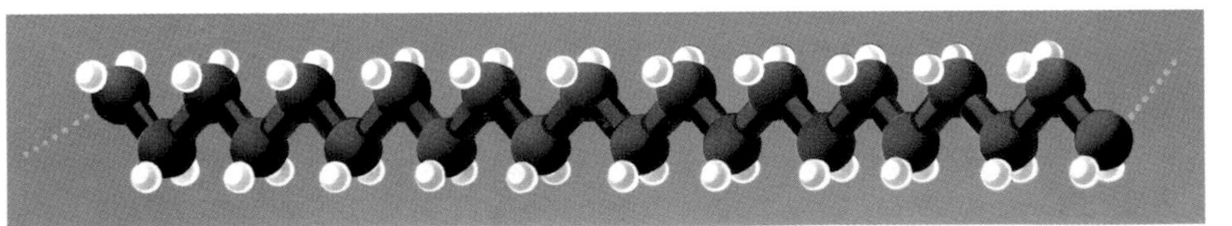

A model of part of a polythene molecule showing repeated units.

Did you know?

- Long chain molecules may contain 100 000 atoms.

1 What is the chemical name for glass? ▲
2 How was oil formed?
3 Describe in your own words how polyethene is made. ▲
4 Polythene bottles can't be used for fizzy drinks. PET bottles can. What does this tell you about the structure of these materials? (The kinetic theory will help you to answer this in terms of the size of carbon dioxide molecules.)
5 Silicon is the second most abundant element on Earth. How do you think this might affect the cost of glass bottles?

Electronics 19

You have had a surprise visit from some friends. You have been busy talking to them. Meanwhile, in the kitchen there is chaos. Several appliances are not working properly.

Make a list of how each is going wrong and then make a guess why.

Out of control

Normally, those machines have something which **senses** that things are going wrong, and which then **controls** them.

The same sort of **control systems** help to keep our bodies working. When we sense that we are too hot, we sweat to cool down. Can you think of any other examples?

Many appliances use *electronic* systems to control them. This section is about the way these systems work to make our lives easier, safer, and more fun.

19.1 Electronics

Q: What have all these in common?

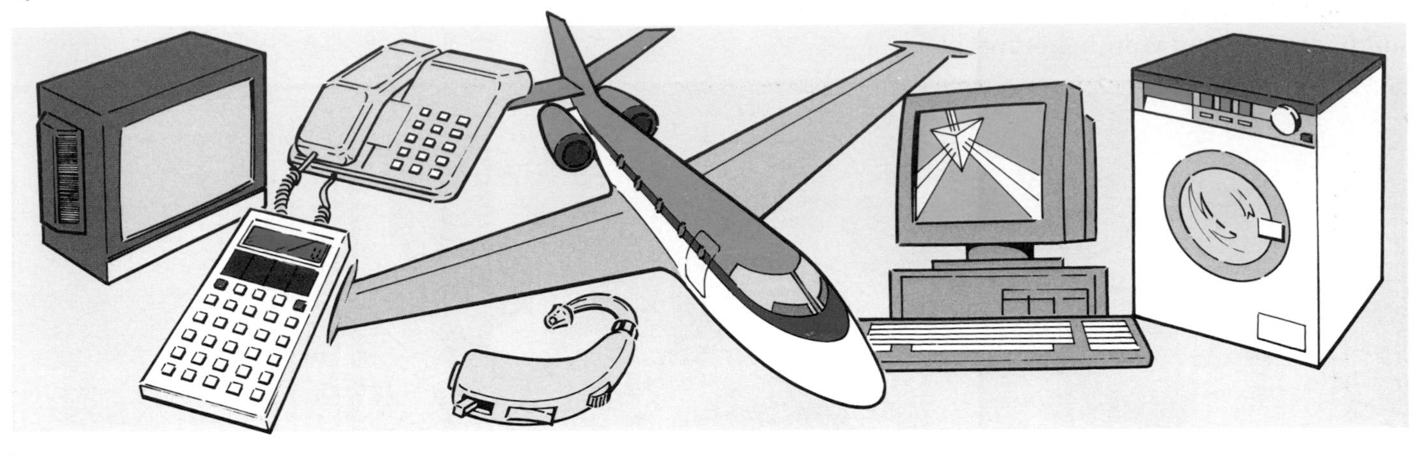

A: They all use **electronics** to do at least part of their job.
You have already learnt about **electricity** which is the energy in **static** or **moving** electric **charge**. Moving electric charge is called a **current**, and it flows in a **circuit**. Circuits are made up from different components.
Electronic circuits are designed so that changes produced in very small currents can be used as signals to make other things happen.
These signals may control other, larger electric currents or may store information.
Many different electronic circuits can be combined to produce systems that can carry out very complicated tasks. Some even control car production lines!
They can react to changes in the same way that people do, and can be designed to 'learn' by experience. They don't learn in the same way as humans, though.

Microelectronics

In microelectronics, this

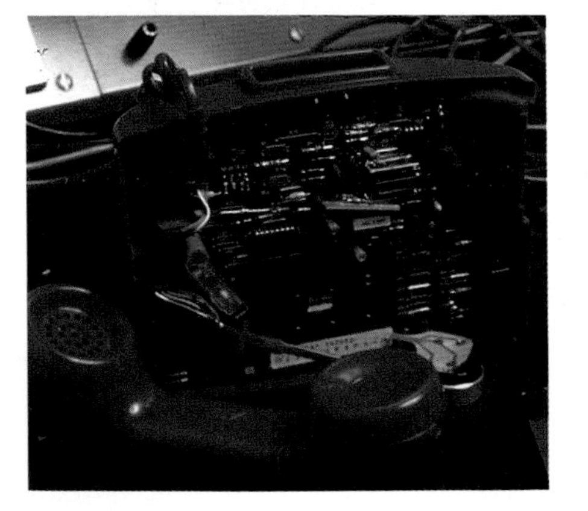

is reduced to this

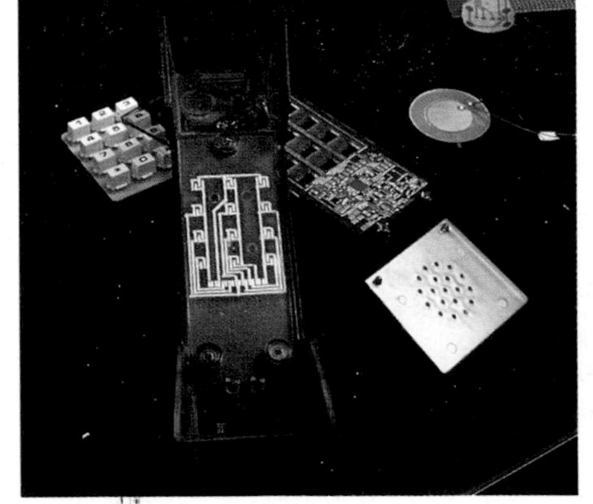

1 Describe in your own words what electricity is. ▲
2 What is special about electronic circuits? ▲
3 What is microelectronics? ▲
4 What job would you most like to invent an electronic device to do?

Did you know?

- By using materials with special properties, known as semiconductors, complicated electronic circuits can be 'drawn' onto tiny areas. Some semiconductors are based on the element *silicon*. These 'micro-circuits' are called silicon chips. When connections are added to the chip, you have an **integrated circuit**. Integrated circuits can be printed onto the plastic casing of appliances.

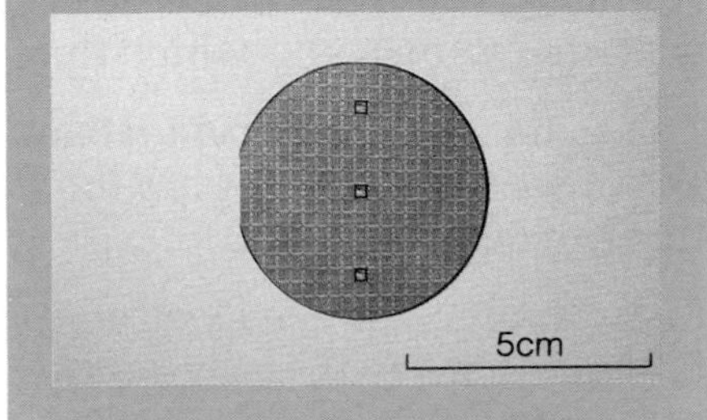

19.1 Revising circuits 1

Going further

Here are some reminders about electric circuits.

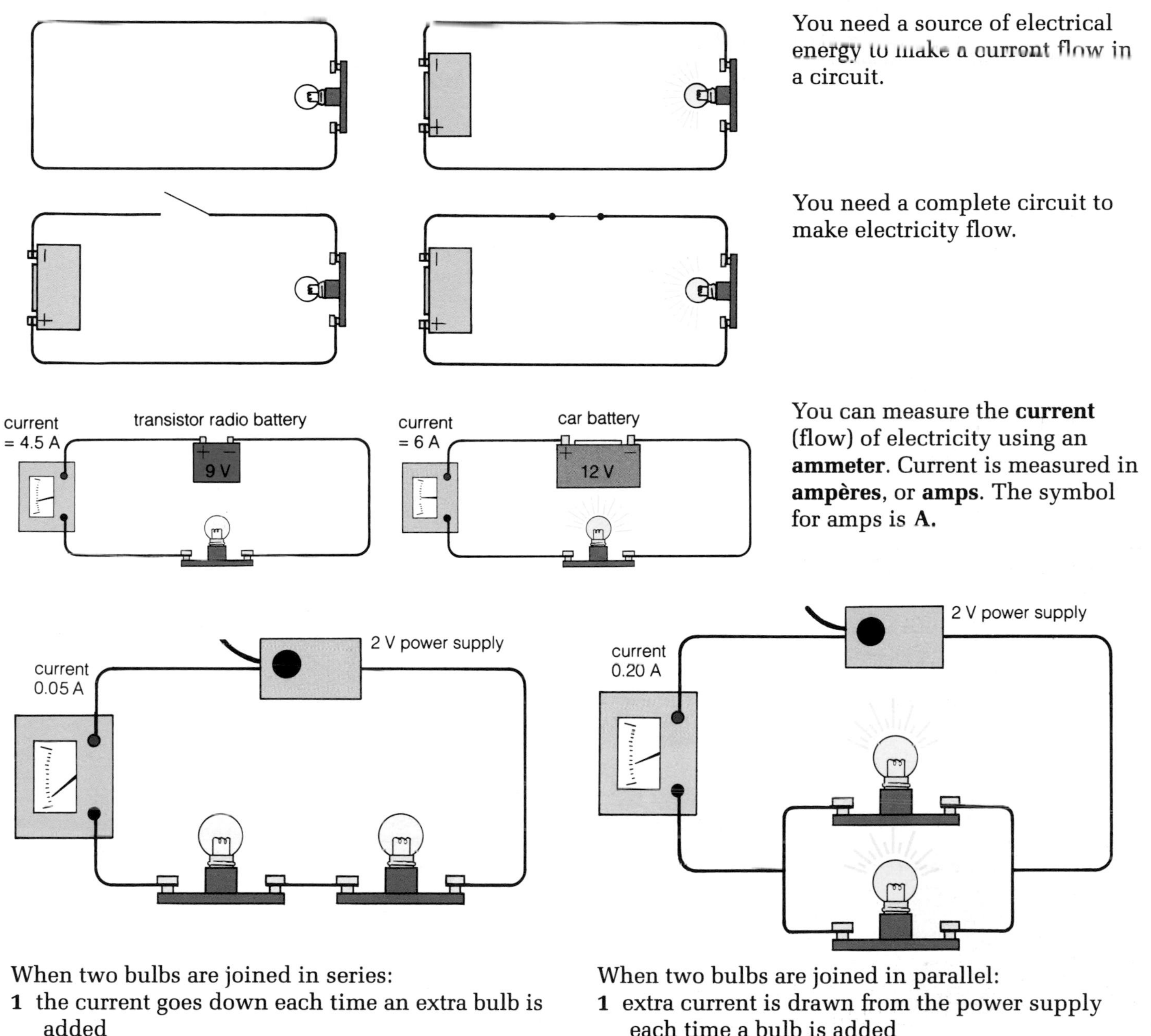

You need a source of electrical energy to make a current flow in a circuit.

You need a complete circuit to make electricity flow.

You can measure the **current** (flow) of electricity using an **ammeter**. Current is measured in **ampères**, or **amps**. The symbol for amps is **A.**

When two bulbs are joined in series:

1 the current goes down each time an extra bulb is added
2 the bulbs get dimmer each time an extra bulb is added
3 all the bulbs go off if there is a gap in the circuit

When two bulbs are joined in parallel:

1 extra current is drawn from the power supply each time a bulb is added
2 the bulbs stay bright even when extra bulbs are added
3 each bulb can go on and off without affecting the others

1 Name one source of electrical energy that you have used in circuits. Where does the energy come from?
2 What happens to the current if a circuit is not complete? ▲
3 Is 'house wiring' arranged in parallel or in series? Why is this useful?
4 What unit is electrical current measured in? ▲

19.1 Revising circuits 2

For the enthusiast

You can measure the 'electrical push' or **potential difference** between two points in a circuit using a **voltmeter**. It is measured in **volts** (symbol **V**).

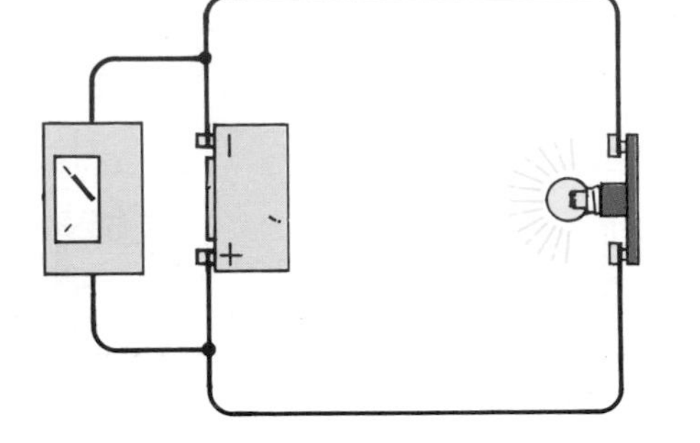

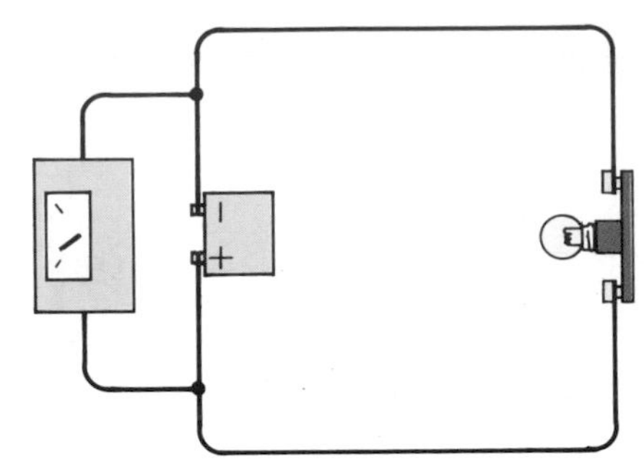

Substances which allow electricity to flow through them are called **conductors**. Other substances do not normally allow current to flow. These are called **insulators**.
Even conductors resist the flow of electricity in a circuit, and so reduce the current. The **resistance** of a component is measured in **ohms** (symbol Ω). When the resistance of a component is high, then a high voltage is needed to make current flow.

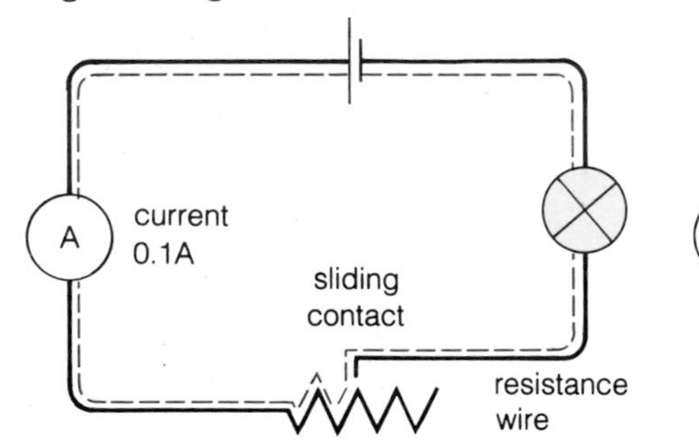

Only a small amount of resistance wire is in this circuit.

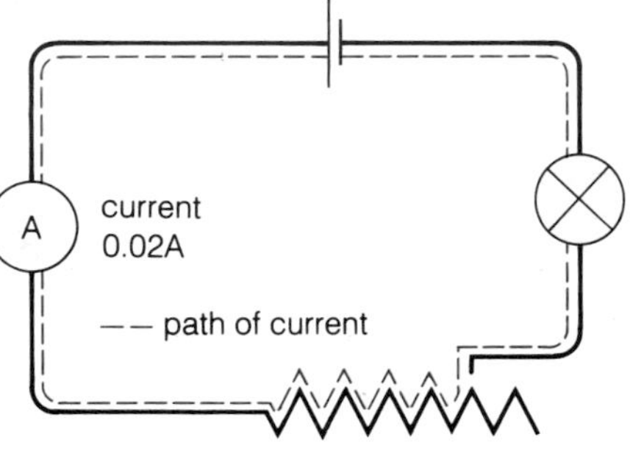

There is more of the resistance wire in this circuit. The circuit has a higher resistance. The current is lower and the bulb is dimmer.

Did you know?

- Insulators, such as air, have a very high resistance. It takes many thousands of volts to make electricity pass through air (in lightning).

Symbols

To make drawing circuits easier, symbols are used to show components:

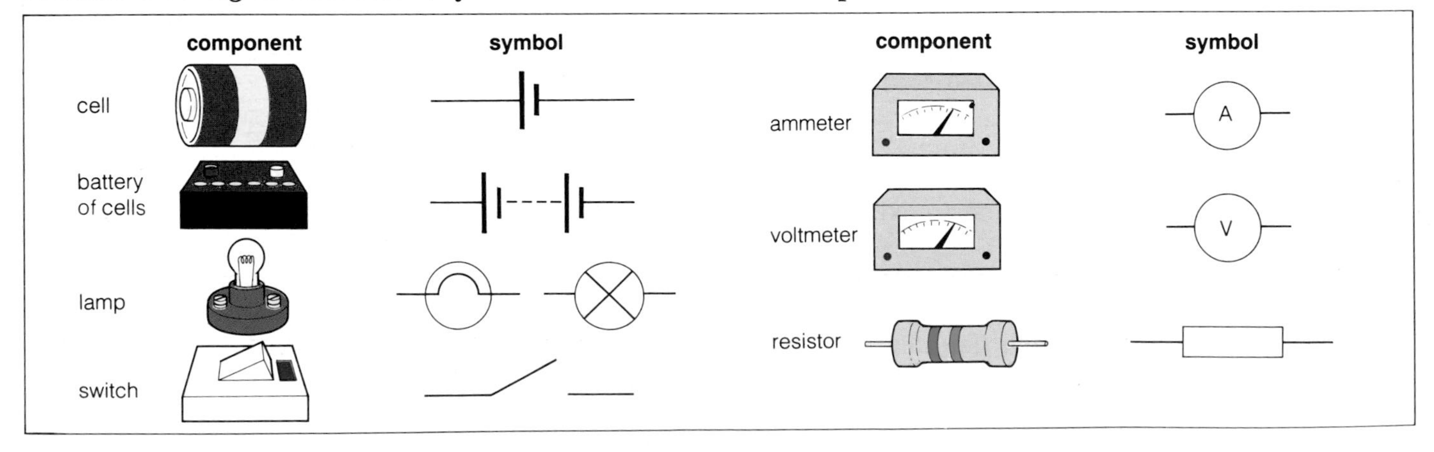

1 What are volts used to measure? ▲
2 Give three examples of a) insulators b) conductors.
3 Use the symbols to draw a circuit containing two cells, a switch, an ammeter, and a lamp. How could you use this circuit to see if a substance was an insulator?
4 What would happen to the current in a circuit if the resistance was decreased and the voltage stayed the same? ▲

19.2 Electronic systems

Starting off

When electronic engineers start to work on a design task, instead of thinking about the hundreds of components that they may use, they look at **electronic systems**.
They look at the behaviour of the system as a whole and the job it has to do.
All electronic systems have three parts: an **input**, a **processor** and an **output**. The processor modifies information received from the input and may make decisions as a result. Some processors have an extra part, a **memory**, which can store information. Looking at what each **part** of the system does helps you to understand how a system works as a whole. Here are three examples of electronic systems

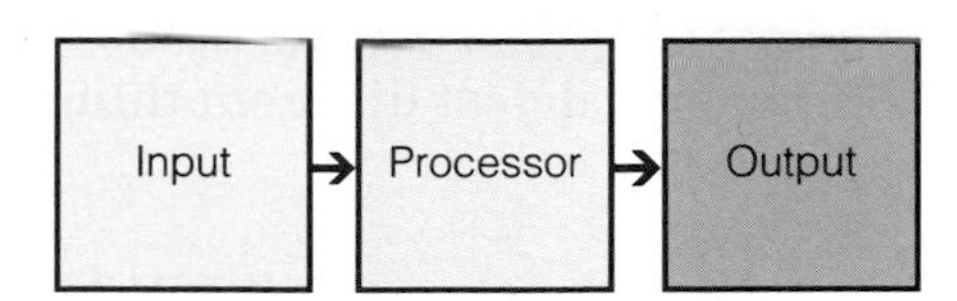

	INPUT	PROCESSOR	OUTPUT
CALCULATOR	**keyboard**	**integrated circuit with memory**	**digital display**
	converts key presses into electrical signals	processes information	displays the results
AUTOMATIC CAMERA	**light meter**	**integrated circuit**	**shutter motor**
	light alters resistance of a light dependent resistor	calculates exposure on the basis of LDR current	shutter opens for the required time
CD PLAYER	**pick up**	**amplifier**	**loudspeaker**
	converts information from the laser light reflected from the disc into electrical signals	makes small electrical signals larger	converts electrical energy into sound energy

1 What are the three parts that all electronic systems have? ▲
2 Explain how a CD player works by describing what each part of the electronic system does. ▲
3 Which part of the electronic system of a tape player would be different from that of a record player?

19.2 The input

Going further 1

Electronic systems are often used to convert changes in their environment to something which is more controllable or measurable. **Input** components detect these environmental changes, which then produce electrical signals in the rest of the system. Different input components detect different things:

Light

Light dependent resistor (LDR) The resistance of a LDR depends on how much light falls on it. In the dark, it has a high resistance, and so only a small amount of current can flow through it.

Solar Cells Convert about 15% of the energy from light to electrical energy. This energy can be used to produce an electrical signal, but is also enough to power things like calculators.

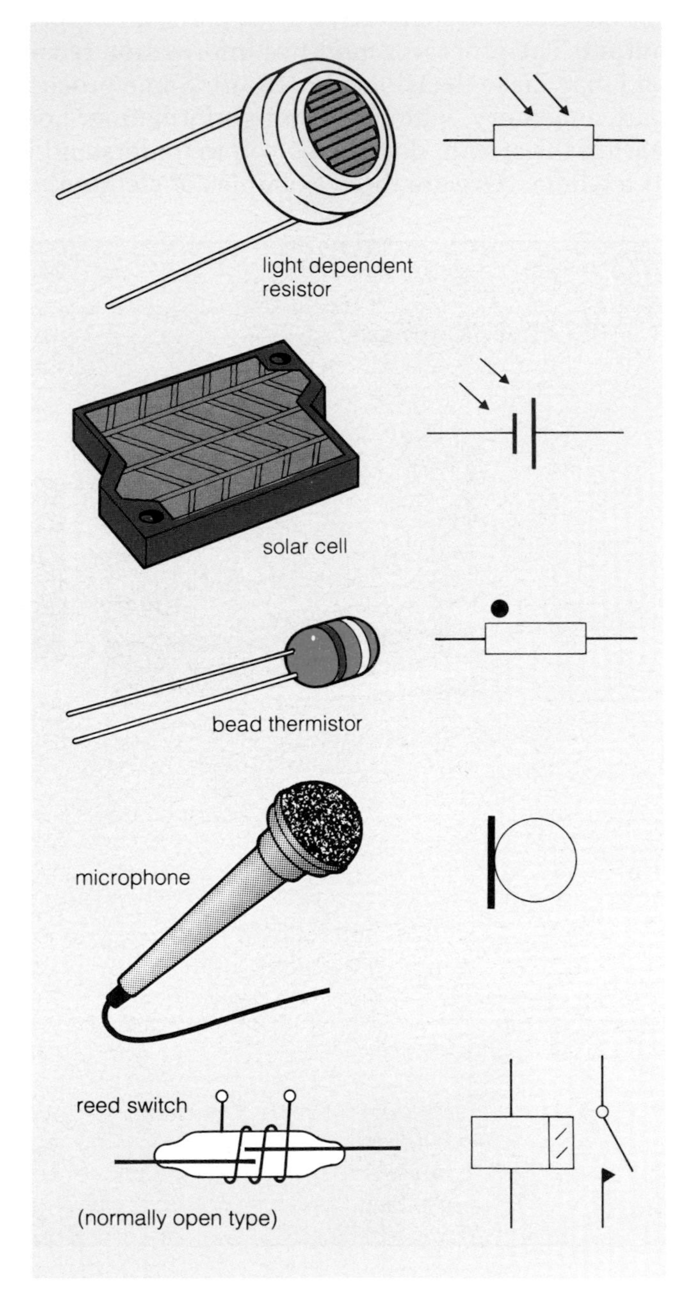

Temperature

Thermistors are like LDRs but respond to heat. The resistance of a thermistor falls when it is heated. This allows a larger current to flow.

Thermocouples convert heat energy to a small amount of electrical energy.

Sound

One type of *microphone* contains a magnet and a coil of wire. Sound waves move the coil along the magnet and this generates a tiny electric current.

Movement

Switches only conduct electricity when they are closed. *Tilt switches* contains a blob of mercury near the switch contacts. When the switch is tilted, mercury flows across the contacts and completes the circuit.

Magnetic field

Reed switch The 'reed' is made of steel, so that the switch can be closed or opened by a magnetic force. This force can be from a permanent magnet or from an electromagnet. Reed switches can be either normally 'open' (the magnet closes the switch) or normally 'closed' (the magnet opens the switch).

Moisture

A moisture sensor consists of two contacts close together. When the contacts are dry, resistance is high. When water lies between the contacts the resistance falls and current flows.

1 What do all input components
   a) detect b) produce? ▲
2 What does a LDR detect? ▲
3 For each of the components described, write how you might use them around the school.

19.2 The output

Input components detect changes in the environment and send electrical signals to the processor. The processor then sends a signal to the output.

Output components receive these electrical signals and react. Different components react in different ways. In designing an electronic system, you need to choose the most useful component for the job.

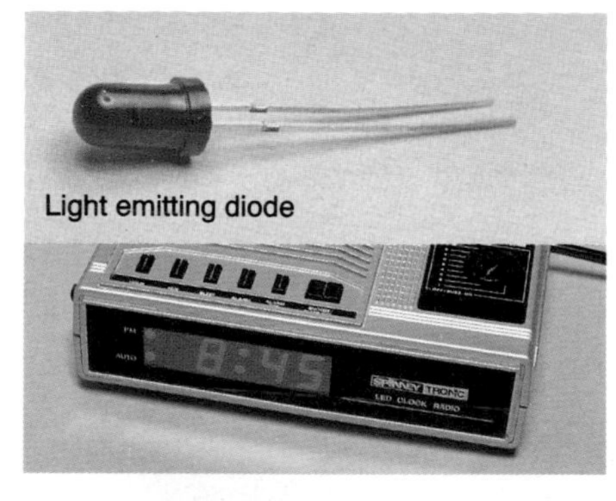

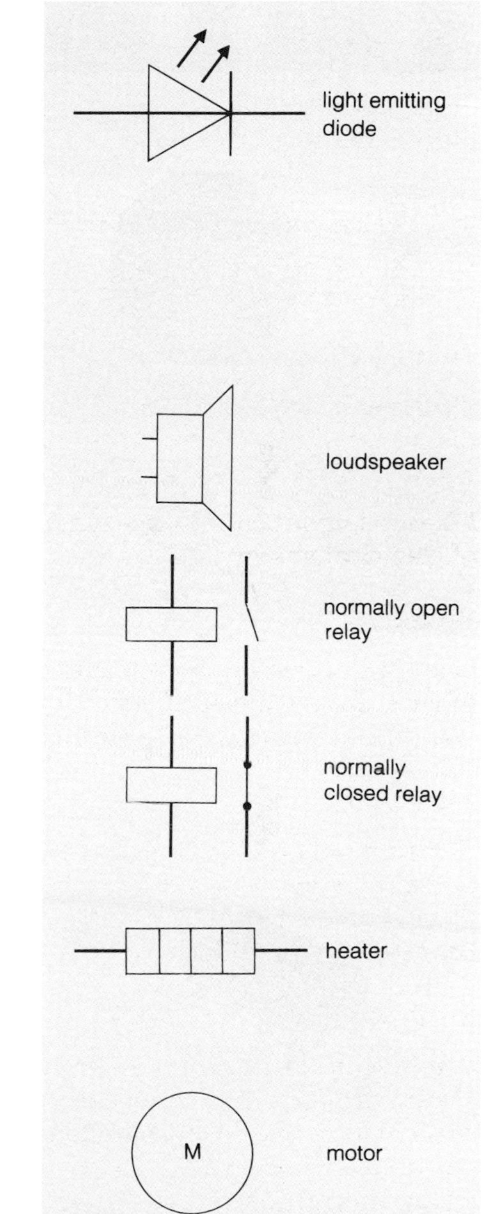

Light

In some calculators and digital clocks, electrical energy is converted into light in a *light emitting diode* (LED). This shines when a current passes through it. Compared to a normal lamp, it works on a smaller current and it produces less heat. LEDs are very useful for indicator lamps.

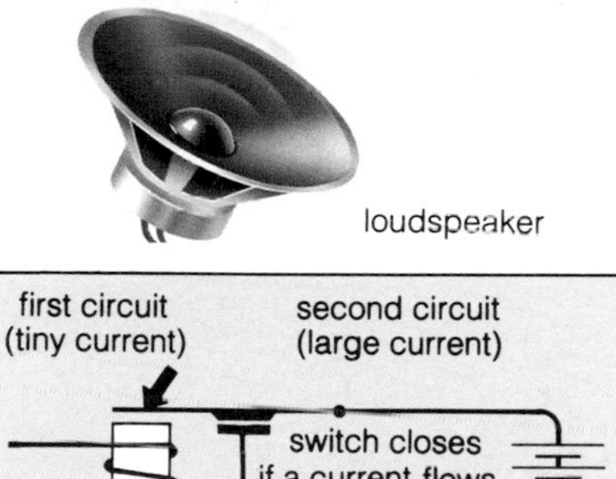

Sound

In *loudspeakers, buzzers*, and *alarm sirens*, electrical energy is converted into sound energy.

first circuit (tiny current)
second circuit (large current)
switch closes if a current flows in first circuit
M
iron core

Relays make a circuit with a small current control a circuit with a large current. When a current flows in the first circuit, the electromagnet attracts the switch contact bar and closes the gap in the second circuit. The second circuit may include a heater or motor.

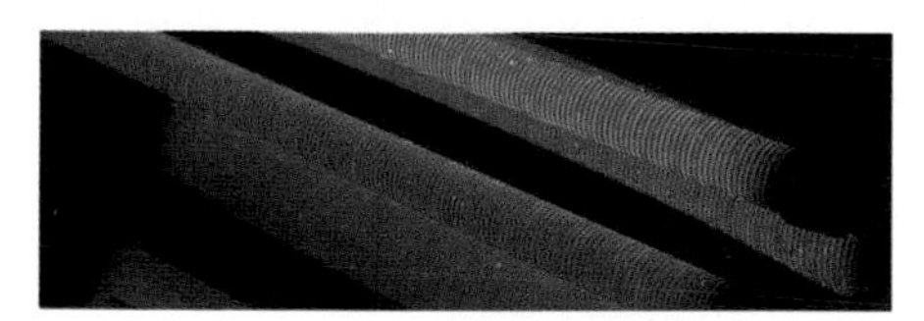

Temperature

The flow of electricity in the relay second circuit may be used to operate a *heater*.

Movement

Electrical *motors* can convert electrical energy in the relay second circuit into movement. This could operate wheels or fans.

1 What do all output devices a) detect
b) produce? ▲

2 What kind of output device could you use
a) to warn you of a fire
b) to let you know that the freezer was switched on? ▲

3 Write a sentence on how each output component could be used around the school.

4 Which input and output components would you combine to
a) warn you that the fridge is not working
b) turn on an outdoor light when it gets dark
c) warn you that someone has broken into your house? ▲

19.2 The processor

For the enthusiast

In between input and output are components which change and control the electronic signals in different ways. They may also perform the input and output job in some electronic systems.

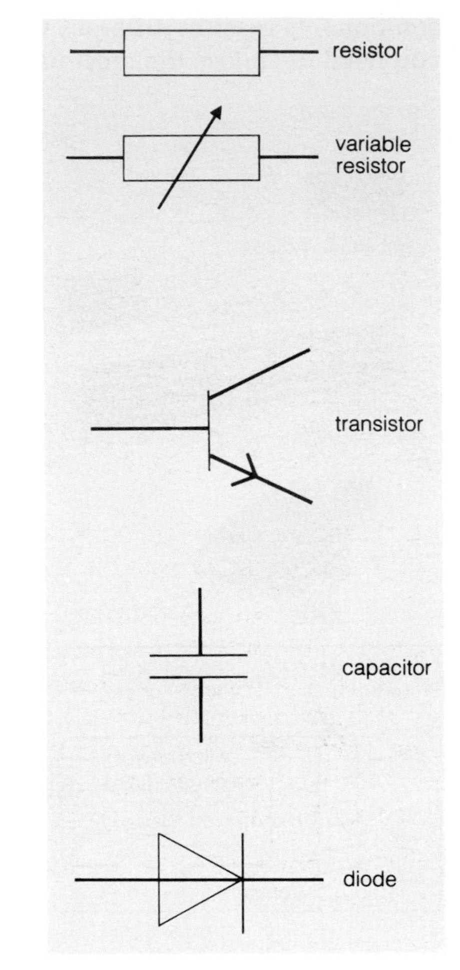

Resistors are used to reduce the current flowing through other components. You can vary the amount of resistance in a circuit by using a variable resistor.

Transistors can be used in two ways. In the first, the pattern of the output current is an 'enlarged' version of the current it receives. It acts as an amplifier. In the second, the transistor acts as a switch. Changes in the current received switch the output on and off. It is much faster than a mechanical switch, and does not wear.

Capacitors are used like 'buckets' to store electrical charges. This is useful if you want a smooth output signal from a variable input signal. In some circuits capacitors and resistors work together to produce time delays.

Diodes only let current through in one direction. This is useful if the input signal is an **alternating current**, like our mains supply, and a **direct current** (one direction only) is needed by the output component.

Integrated circuits. Modern electronic systems are very complicated. Nowadays, hundreds of resistors, diodes, capacitors, and transistors can be produced on one tiny chip of silicon the size of a pin head. These tiny packages of components, with connections fitted to them are called integrated circuits (ICs).

When you use an electronic system, you need to combine the right input, processor and output components for the job.

1 Why might you want to design a system with a time delay in it?
2 List the input, output and processor components that you could use in a circuit to do the job you described in **1**.
3 List the components that you would need in an electronic system designed to turn an *alternating current* from the mains into a smooth *direct current*.

Did you know?

- In 1904 Ambrose Fleming invented the first electronic component, a rectifier valve. This has been replaced by a much smaller 'junction' diode in modern circuits.

19.3 Analogue or digital?

Starting off

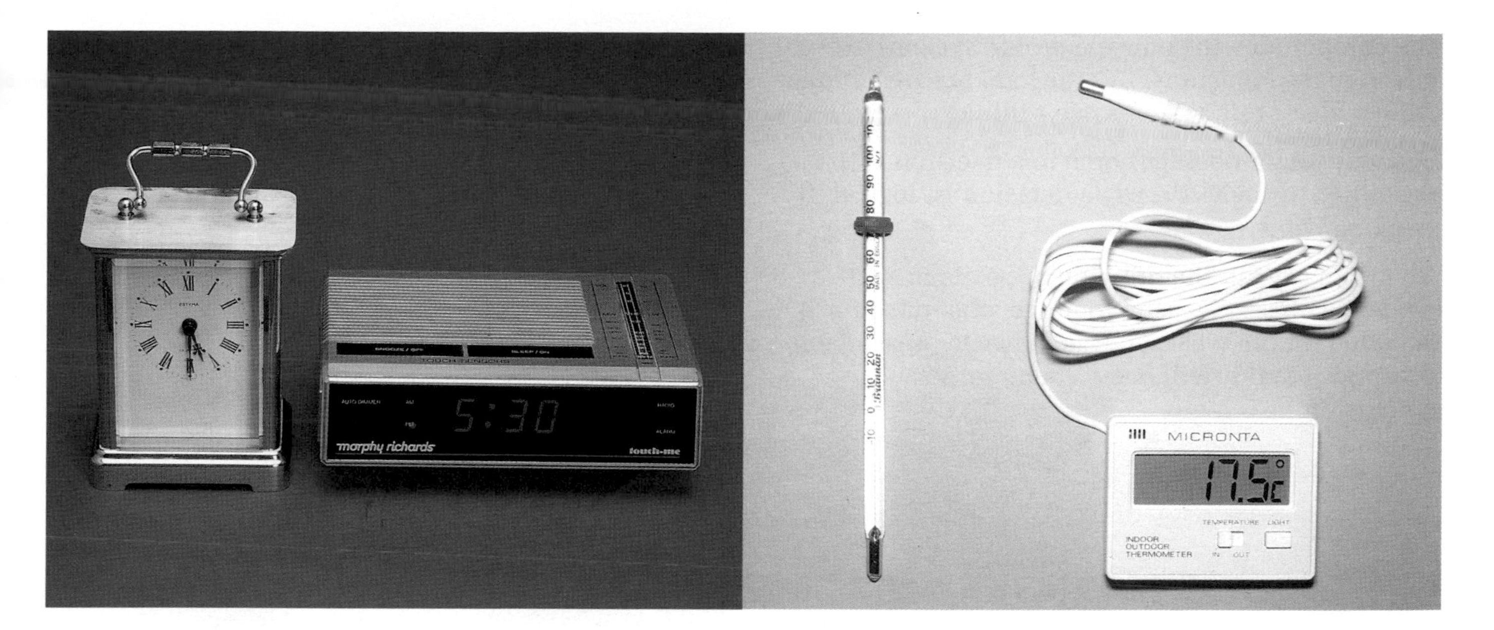

You should get the same readings whether you are using *analogue* or *digital* equipment, so what is the difference between them?

Digital equipment nearly always uses numbers to show readings. Analogue equipment uses the position on a scale. Analogue equipment shows a continuous change in readings. You can watch a liquid thread gradually creep up a thermometer. You have to judge when the liquid has reached a certain temperature. A **digital** thermometer moves in definite steps: 29.9°, 30.0°, 30.1°. It will spend some time on each reading before moving onto the next.

In most digital equipment quantities are turned into numbers during recording.

Did you know?

- The numbers from 0 to 9 are called **digits**. This is also another word for your fingers and toes. The connection is the way we learn to count!
- Music on compact disc is recorded **digitally**. Sound signals are converted to numbers. A CD player translates these digital messages into analogue signals, which a loudspeaker can convert into sound.
- Information on a record and on a CD is stored on a spiral track. On a record, the stylus starts at the outside and follows the track inwards. On a CD, the laser beam starts from the inside and follows the track outwards.

1 Explain the difference between analogue and digital equipment. ▲
2 Digital music recordings contain separate 'packets' of information. Why don't you hear these as separate 'bursts' of sound when they are being played? ▲
3 Make lists of analogue and digital equipment that you have at home.

19.3 Logic

Going further

The components in some electronic systems are able to make decisions. But they are not like human beings. Electronic systems can't think!

One way human beings come to a decision is to use reasoning, or **logic**. We say a decision is 'logical' if it seems sensible.

What Tim says in the drawing opposite is not **logical**. He has no reason to judge what day it will be, based on the weather. It is, however, reasonable to suppose that it will snow soon. Tony's conclusion uses logic.

The logic of electronic processors is much more basic than this. They make decisions between only *two* alternatives at any one time. These are called **binary** decisions.

The information fed into a processor is in **digital** form. Each processor decision is based on a choice between the binary numbers '0' and '1'.

Binary codes and electronic systems

Let's now look at how the conditions in an electric circuit can be linked to binary numbers. You could ask, "Does a current flow?" or "Is there a voltage?" The answers would be either "yes" or "no". That can be converted to digital information by making yes = 1 and no = 0. The circuit is producing binary information.

In an electronic system, the processor receives information from the input device and passes information to an output device. Look at the table below to see how current and voltage in these circuits can be shown as binary states "0" and "1". You will need to refer to this information again.

binary code	simple choice	current	voltage	switch	LDR	thermistor	moisture meter	buzzer	LED
0	no	low/zero	low/zero	off	dark	cold	dry	silent	unlit
1	yes	high	high	on	light	hot	wet	buzzes	lit

In terms of logic, the descriptions of the states in the table above are called **logic 0** and **logic 1**.

1 Write down an example of a logical decision you could make.
2 Describe how electrical signals are shown as binary information. ▲

19.3 Logic gates

For the enthusiast

The decision-makers in electronic systems are called logic gates. They are called gates because they have to receive the right input signals before they let output information through.

There are five sorts of logic gate, each with one or two inputs and one output. The names of the five gates are **NOT**, **AND**, **OR**, **NAND** and **NOR**. Each type of gate has a different symbol. Symbols can be joined to other devices in a block diagram to show an electronic system.

For each gate, the input, or inputs are on the left of the symbol. The output is on the right. Each input and output can be either high (logic 1) or low (logic 0).

The name of the gate gives you a clue to which input signals are required before a high (logic 1) signal is passed to the output. In the table below, the **truth table** on the right shows the electrical state at the inputs and output in each case (remember : 0 means low voltage and 1 means a high voltage). In a truth table, all possible inputs are included.

Did you know?

- Each symbol which says something *negative* has a blob on its 'nose'.

Gate	Input	output	Description	Truth table
NOT			Output is high (logic 1) when input is **not** high	input output 0 1 1 0
AND	A B		Output is high (logic 1) when input A **and** input B are high	Input A input B output 0 0 0 0 1 0 1 0 0 1 1 1
OR	A B		Output is high (logic 1) when input A **or** input B **or** both are high	input A input B output 0 0 0 0 1 1 1 0 1 1 1 1
NAND	A B		NAND = not AND. Output is high (logic 1) when input A **and** input B are **not** high	input A input B output 0 0 1 0 1 1 1 0 1 1 1 0
NOR	A B		NOR = not OR. Output is high (logic 1) when neither input A **nor** input B is high	input A input B output 0 0 1 0 1 0 1 0 0 1 1 0

The next section shows how these gates can be used in practice.

1 Describe in your own words what a logic gate is and how it can be used to make decisions in electronic systems. ▲

2 For each of these gates write down what you think the logic state of the unlabelled connection is:

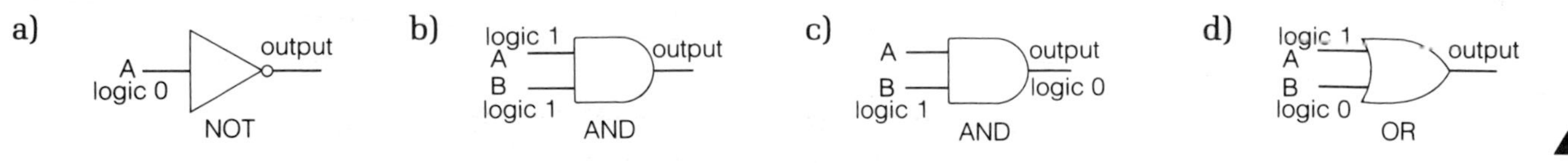

▲

19.4 Using logic gates 1

NOT gates

A NOT gate has one input (on the left) and one output (on the right). When the input is high (logic 1), its output is low (logic 0). When the input is low (logic 0), the output is high (logic 1).

For example:

When the kettle is switched on the light is out. When the kettle boils and the heating circuit switches off, the light goes on.

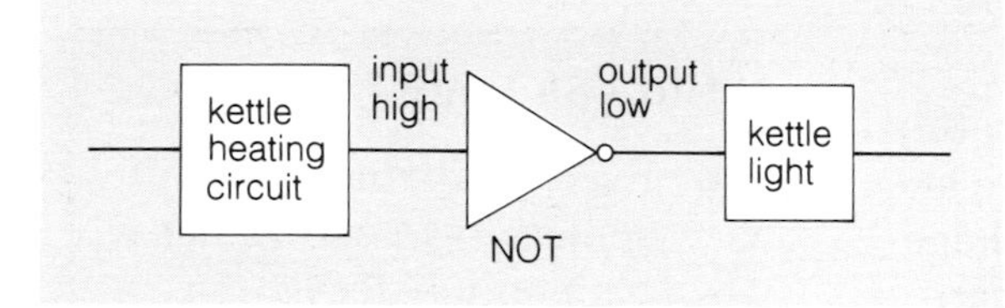

Truth Table:

Input	Output
0	1
1	0

AND gates

An AND gate has two inputs and one output. The output is high (logic 1) only if both inputs are high (logic 1).

For example:

A gardener may decide that it is best for an automatic plant sprinkler to operate only when it is dark and cool. She could use a system like this:

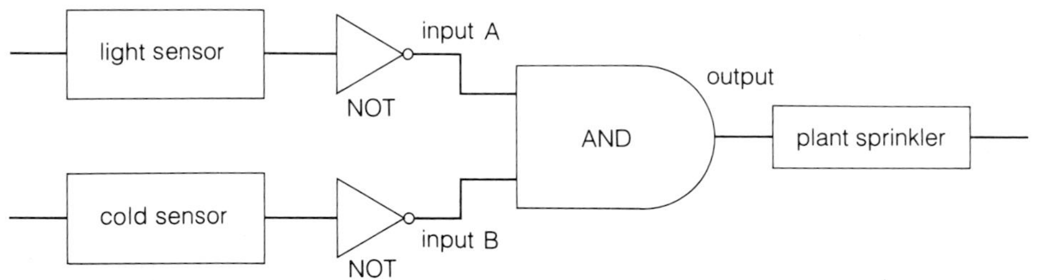

Truth Table:

Inputs A	Inputs B	Output
0	0	0
0	1	0
1	0	0
1	1	1

1 Draw a truth table for
a) a NOT gate b) an AND gate ▲

2 An AND gate could be used in a washing machine to make sure that the heater comes on only after the door is closed *and* the machine is filled with water. What two input devices, and what output device would be connected to the gate? (Remember that the output current from the gate will be small, but that the heater will need a large current.)

3 If you leave a car with the headlights on for day, the battery normally becomes 'flat', and hasn't enough power to start the car again. Use the items here to draw a system that bleeps if you leave the lights on after turning off the car engine.

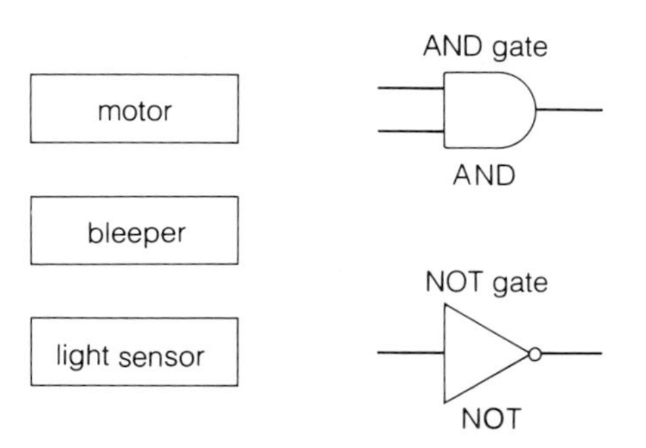

19.4 Using logic gates 2

OR gates

Like an AND gate, an OR gate has two inputs and one output. The output is high (logic 1) if both inputs are high, or if only one input is high.

For example:

If either the smoke detector or the temperature detector are on, or both, the alarm will ring.

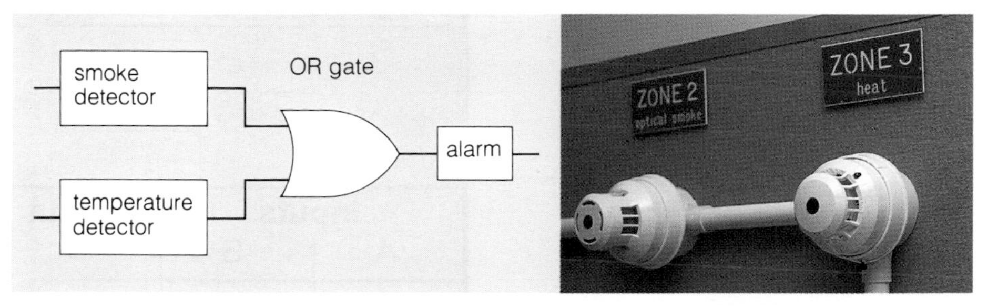

Truth Table:

Inputs A	Inputs B	Output
0	0	0
0	1	1
1	0	1
1	1	1

1 Draw the block diagram for a burglar alarm which will operate when a window is opened or when a light is switched on. Label it.

For **2**, **3** and **4** you may choose to use NOT, AND, or OR gates. For each case, first decide which gate to use. Then, using the correct symbols, draw a block diagram for the electronic system, and label the input and output devices.

2 This machine is dangerous! To start it, two switches have to be pressed at the same time. This makes sure that the operator's hands can't be trapped.

3 The motor of this fridge-freezer will switch on when either the freezer compartment or the fridge compartment get too warm.

4 These yachts have an automatic anchor light for when nobody is on board. In the evening, when it gets dark, the light goes on. In the morning, it turns off again.

19.4 NAND and NOR gates

NAND gate

The symbol for a NAND gate is shown here. You can see it has two inputs (on the left hand side) and one output (on the right).

The output is high (logic 1) only if at least one input is low (logic 0). This is the equivalent of having an AND gate followed by a NOT gate (NAND = AND NOT).

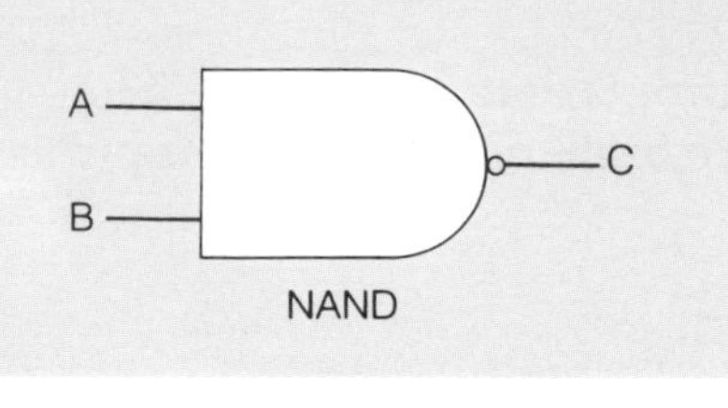

Truth Table:

Inputs		Output
A	B	
0	0	1
0	1	1
1	0	1
1	1	0

NOR gate

The symbol for an NOR gate is shown here. You can see it has two inputs (on the left hand side) and one output (on the right).

The output for a NOR gate is only high (logic 1) if both inputs are low (logic 0). This is the equivalent of having an OR gate followed by a NOT gate (NOR = OR NOT).

Truth Table:

Inputs		Output
A	B	
0	0	1
0	1	0
1	0	0
1	1	0

Did you know?

- 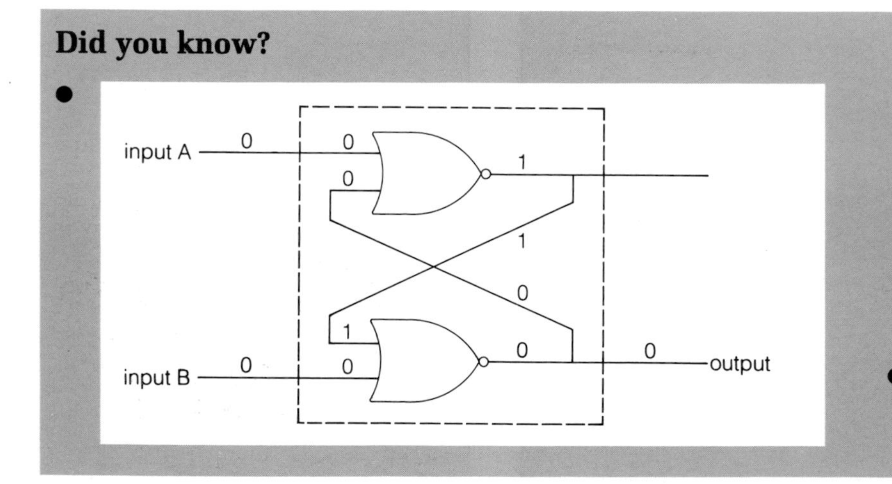

  This circuit is called a **bistable**. It has two stable states. Both inputs are normally logic 0. The circuit can detect short changes in logic at either input. If input A was last raised to logic 1, the output stays at logic 1. If input B was last raised to logic 1, the output stays at logic 0.
- Groups of bistables can be used as *memories* and as *counters*.

1 a) Check that a NAND gate is equivalent to having an AND gate followed by a NOT gate.
b) Check that a NOR gate is equivalent to having an OR gate followed by a NOT gate.
Hint: Write out a truth table for the first gate, and use the output logic from this as the input logic for the NOT gate. ▲

2 Draw the block diagram of a system which will start a heater fan when a room becomes cold and dark.

3 Design and draw the block diagram for a hot water cylinder protection system. This should shut the heater off
a) if the water starts boiling (i.e. gets too hot) or
b) if the water level falls too low.
You can use any combination of gates.

4 By writing truth tables for each stage, investigate the effect of putting a NOT gate in one of the inputs of an AND gate. ▲

Energy and environment 20

What is life?

Can you spot the differences?

Here are some features of living things. They:
- feed
- grow
- reproduce
- excrete waste substances
- can sense things in their surroundings and respond
- can move

For life to exist these **resources** need to available:
- water
- oxygen
- carbon dioxide
- light
- nutrients
- warmth

Plants use energy from the Sun to grow. Plants provide food for animals. Animals and plants obtain the things that they need from their surroundings or **environment**.

Different types of environment, such as woodland, coast or rivers, are called **habitats**.

Animals and plants are **adapted** to living in a particular sort of natural habitat. They have features which enable them to obtain everything that they need to grow and reproduce from the habitat.

Humans have also become adapted to certain ways of life. We use large amounts of energy and materials to make life comfortable. But building, industry, farming, and mining may change natural habitats. Animals and plants may not be so well adapted to the new conditions. And some of our activities may be damaging our *whole* environment permanently.

Ecology is the study of animals and plants, and their habitats.

Ecologists make **surveys** of the plants and animals in the habitat, and measure conditions such as temperature, light, and acidity in the soil. An **ecosystem** is a collection of plants and animals and their environment.

This chapter starts with a brief look at ecology, and how changes can be made to ecosystems to produce our food. It then goes on to look at the effects which producing energy and materials can have on the environment.

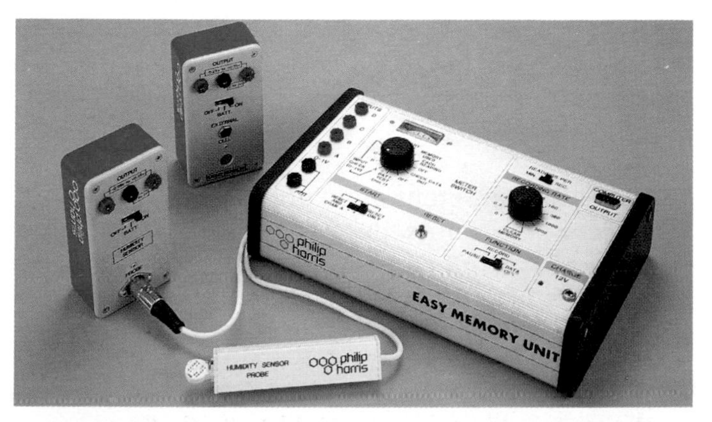

A single instrument can be used to collect data on the conditions in an environment. This environmental probe uses **electronic** devices to measure temperature, pH, and humidity, for example.

20.1 Plants 1

Starting off 1

Plants are vital to life on Earth because they can use energy from the Sun to grow. They are at the beginning of all **food chains.** Without plants, no animals could exist. Plants are called the **producers** in a food chain, and animals are called **consumers**. There are about 380,000,000 known **species**, or types of plant.

A food chain.

You have already looked at the part that flowers play in plant reproduction. However, many plants reproduce in other ways:

Mosses and Liverworts These are usually small plants. They don't have seeds but use tiny **spores** to reproduce. They are found in damp places.

Ferns are large plants. They have no seeds, but produce spores on the back of their leaves.

Conifers produce seeds. They have male and female cones. The male cones produce pollen which is carried by the wind to female cones. When these are fertilised, seeds develop.

1 Name four different plant groups. ▲
2 In your own words, describe how conifers reproduce. What else do you know about conifers?
3 Why are plants such an important part of ecosystems? ▲
4 What sort of habitats do mosses grow in? ▲
5 Complete these sentences using the words given here:
producers *consumers* *food chain*
Plants use energy in sunlight to grow. They are the _________ and the beginning of every __________. Animals are ________. ▲
6 What happens to plants if you leave them in a dark cupboard?

Did you know?

- There is a fern with 'leaves' only one cell thick. This is the kidney fern, found in New Zealand.

20.1 Woodland competition

Starting off 2

We know that all plants need light to grow and reproduce. In a woodland, it may be too dark beneath the trees for small plants to grow. They cannot *compete* with the trees for light.

If the trees are **evergreen** (leaves all year round) smaller plants can only grow if a tree falls over. A light area, or clearing, is left until another tree grows in its place.

If the trees are **deciduous** (lose their leaves over the winter), there is a chance for smaller plants to grow before the leaves appear in Spring. They do not have to compete with trees for light. Once it is warm enough in early Spring, many small plants appear on the woodland floor.

They need to flower and reproduce before the trees take over. They have **adapted** to survive alongside the trees by completing their life-cycle in early Spring.

Evergreen woodland.

Deciduous woodland.

A horseshoe bat. Find out how it uses echoes to 'see' at night.

All animals and plants compete for resources in their habitat. Those which are best **adapted** to their surroundings will survive.

Animals may be adapted to changes which happen every day. Many animals are **nocturnal**. They feed and move around at night, and are not active during the day. Others are active during the day (they are **diurnal**).

Animals adapted to surviving at night can avoid diurnal predators, and do not compete with diurnal animals for food.

1 In your own words describe what nocturnal and diurnal mean. ▲
2 Apart from light, what other resources do plants have to compete for?
3 In your own words, describe what **deciduous** and **evergreen** mean. ▲
4 List the **nocturnal** animals that you can think of.
5 List three more examples of the way plants or animals are adapted to seasonal changes.

Did you know?

- Many animals and plants are sensitive to the length of daylight. Changes in daylength trigger them to reproduce.

20.1 Plants 2

Going further 1

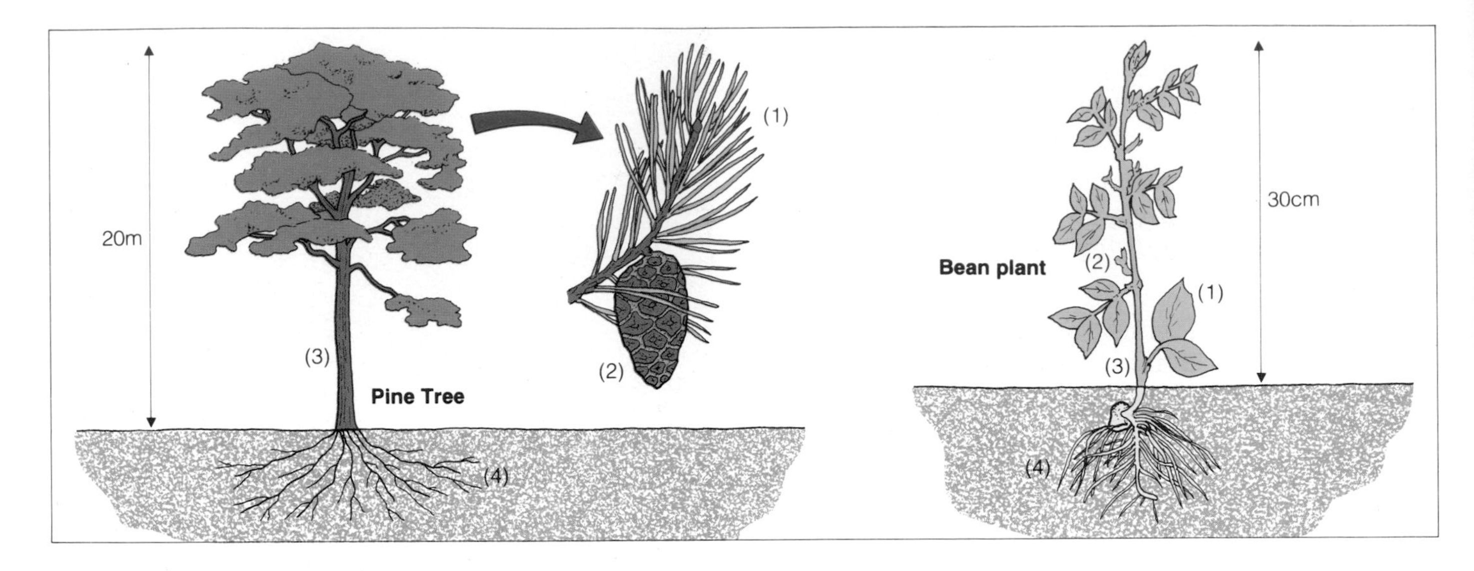

A bean plant may look very different from a tree, but the same basic processes are at work in both.

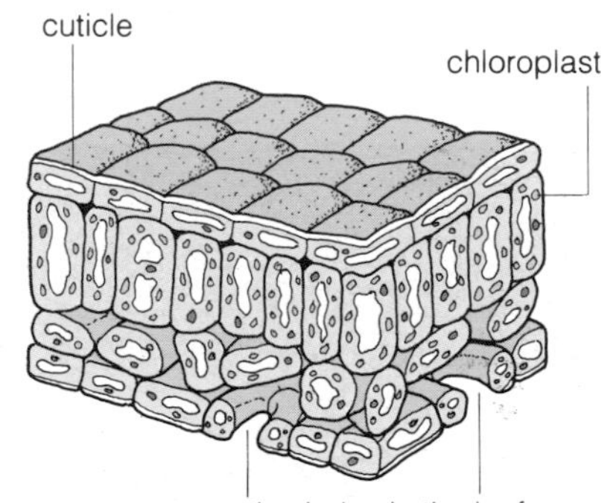

stomata are tiny holes in the leaf. Oxygen and carbon dioxide pass through these holes.

1 Pine needles and bean leaves both trap the Sun's energy and convert it into **carbohydrate** in a process called **photosynthesis**.

This magnified slice of a leaf shows exactly where photosynthesis happens. Chloroplasts contain **chlorophyll** which enables this reaction to take place:

$$\text{Water} + \text{carbon dioxide from the air} \xrightarrow[\text{sunlight}]{\text{chlorophyll}} \text{oxygen} + \text{glucose}$$

Plants also respire all the time like every other living thing to obtain energy to grow. *Remember:*

glucose + oxygen → carbon dioxide + water + energy.

So in the dark, when no photosynthesis takes place, the leaves take in oxygen and give out carbon dioxide.

2 Reproduction takes place in the pine cones and the bean flower, and seeds are produced.

3 The leaves must be high enough to reach enough sunlight. Flowers and cones need to be raised above the ground so that pollen can be transported from or to them, either by wind or insects. Seeds also need to be raised above ground level so that they fall over a wide area.

In bean plants the stem gives this support, and it can bend towards the sunlight. Although the pine's trunk cannot bend, it is **woody** and so can support much taller and wider growth. The plant material which gives this solid structure is called **lignin**.

4 The roots of both plants must spread out far into the soil. They take in water and dissolved minerals, such as nitrate and phosphate, from the soil. These are vital to a plant's growth and survival.

Roots also provide an anchor into the soil for plants, and stop the wind blowing them over.

Did you know?

- Plants are green because they absorb red and blue light for photosynthesis, and reflect back the green light.

1 In your own words, describe where photosynthesis takes place in plants. ▲

2 Make a list of three functions of plant roots. ▲

3 What are the differences between a bean plant stem and a pine tree trunk. ▲

4 Make a list of what gases enter and leave a plant:
a) in the day b) at night ▲

20.1 Population changes

Going further 2

The human world population in 1987 was 5000 million. The population of the United Kingdom is about 57 million. The population of slugs in my garden is 50. The population of crocuses in the park is 2500.

A population is the number of individuals of a particular species in an area. This area can be large or small.

The world human population was about the same (or stable) for hundreds of years. Since the beginning of the last century it has risen sharply. Reasons for this rise include improved medical treatment, and better housing and nutrition.

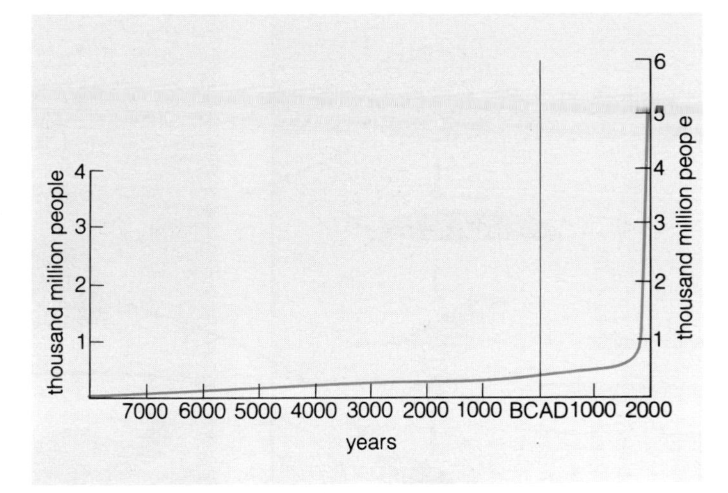

Animals and plant populations

The numbers of animals and plants in an area varies over time, and populations can be completely wiped out. If the world population of an animal or plant falls so much that none are left, the species is **extinct**. Nothing will bring it back.

The size of a population is affected by:

- the amount of food, water and shelter available to each individual
- the number of individuals killed by predators or disease

Of course, for **predators**, which feed on other animals, success depends on the population of **prey** in the area.

Predator and prey in the garden

Aphids (or greenfly) feed on garden plants, especially roses. Ladybirds feed on aphids. When the population of aphids goes up, so does the ladybird population after a short delay. The ladybird population depends on the aphid population.

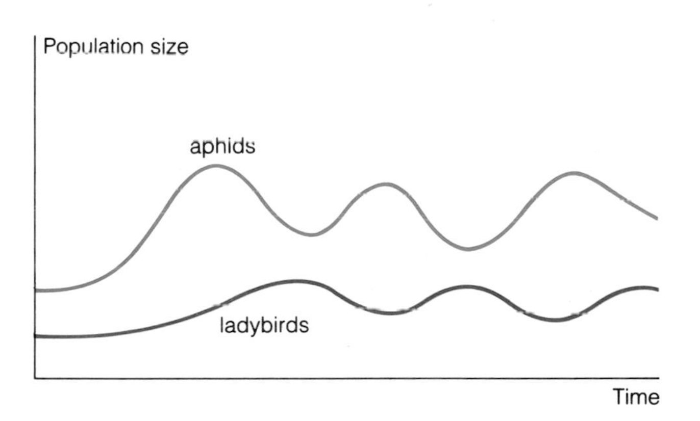

Did you know?

- This bird, the dodo, lived on the island of Mauritius until about 300 years ago. People hunted it so much that it became extinct. Many other species are in danger of extinction today. They are recorded by the International Union for the Conservation of Nature in *Red Data Books*.

1. What is the population of your nearest town or village?
2. Describe what a population is in your own words. ▲
3. Describe how the human world population has changed over the years, and give reasons for the change. ▲
4. Why is the ladybird population smaller than the aphid population in the example? ▲

20.1 The energy chain

For the enthusiast

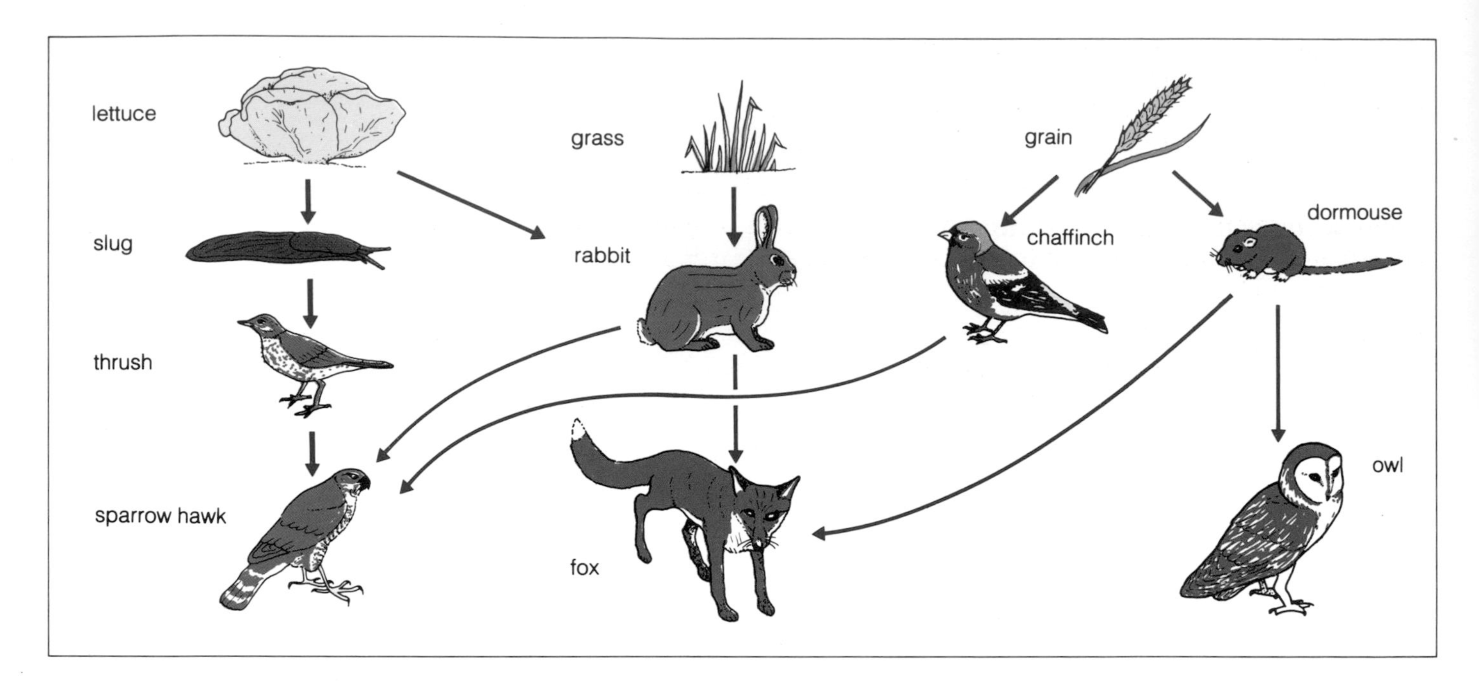

You have looked at examples of food chains and webs. At the beginning of any food chain is a plant using sunlight to convert water and carbon dioxide into carbohydrate. Animals (**consumers**) then feed on plants and use the carbohydrate for energy to grow and move. These animals in turn may be eaten by other animals and so energy passes along the food chain.

However, at each stage in the food chain energy is lost. When a rabbit eats a lettuce only a small part of the energy is 'saved' as new growth. Most is used to keep the rabbit alive, and some is just not absorbed from the lettuce during digestion.

The same 'waste' happens at other stages in the food chain.

Because of this waste, the number of animals at each stage in the chain decreases.

A **pyramid of numbers** shows this:

Another example is shown here.

This is not a pyramid, because there is only one rose bush at the base.

A better way of showing how energy passes along the food chain is a **pyramid of biomass**.

This shows the amount, or *mass*, of plant or animal material that is eaten at each stage.

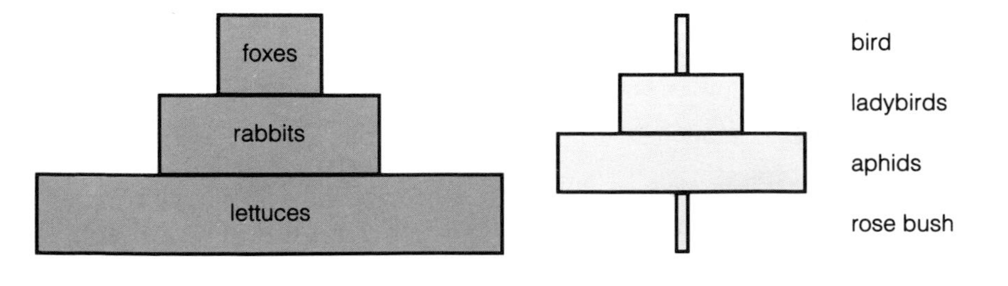

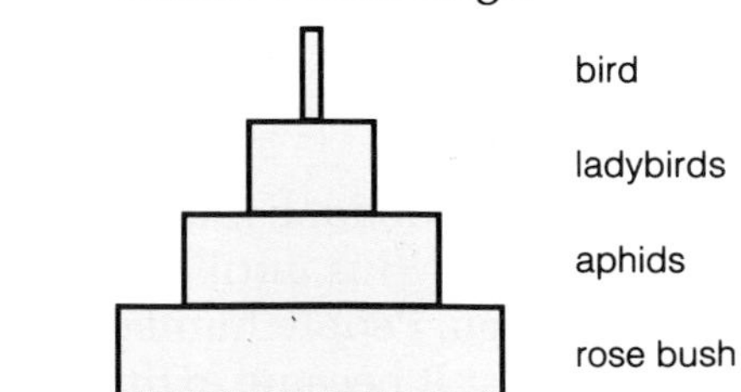

1 What does a pyramid of numbers show? ▲
2 Why do you think that the length of food chains is usually no more than four animals?
3 What does a pyramid of biomass show? ▲
4 Draw what you think a pyramid of biomass would look like for this food chain:
grain → dormouse → owl

20.2 Our need for food

Starting off

Ten thousand years ago, less than 10 million humans lived on Earth. In 1950, the world population was about 2500 million. By 1987, in only 37 years, that figure had doubled to 5000 million. By the time you retire, the population may have doubled again!

Even now, we read about famine in certain countries. If we can't produce enough food in the 1990s, will you be going hungry in the 2030s? It is unlikely. Already, the technology exists to produce increasing amounts of food:

- more land can be made available for food production.
- more food can be produced from each hectare of land.

Clearing the land The obvious way to make more land for farming is to clear away old forests. There are dangers in this. Tree roots hold moisture in the soil. When trees are removed, the topsoil dries out. When it rains, the water runs off the surface and carries the topsoil with it. This is called soil erosion.

Soil erosion.

Boosting production There are several ways in which food production can be increased. These include:

- developing new breeds of plants and animals.
- increasing mechanisation,
- improving water supplies,
- using artificial fertilisers,
- fighting pests which eat crops.

New breeds

New strains of animals and plants are produced by cross-breeding. By selecting the parents for their best qualities, the new breeds can be 'designed' to give increased productivity, or resistance to disease.

This field produces twice as much grain as it did forty years ago.

This cow produces twice as much milk as the average cow did forty years ago.

1 What are two possible ways of increasing food production? ▲
2 Why can't crops be grown on eroded land?
3 In what ways can farm mechanisation increase productivity?
4 **Try to find out** how a British farmer might provide extra water during a dry spell. What effect might this have on the environment?

Did you know?

- In Great Britain, two-thirds of all agricultural land has grass on it. We can't eat the grass, but sheep and cattle convert it into meat and milk!

20.2 Natural recycling

Going further 1

It is annoying when fresh food goes mouldy or rotten before we have had a chance to eat it. Bacteria and fungus start to grow and break down food if it is not preserved properly. But imagine if this process of **decay** never happened. Our food would last longer, but so would dead animals and plants in nature. Not only would this be very untidy, but all the useful minerals and carbon would be locked away permanently in their bodies. The natural process of decay means that these valuable building blocks of life are continually recycled and made available.

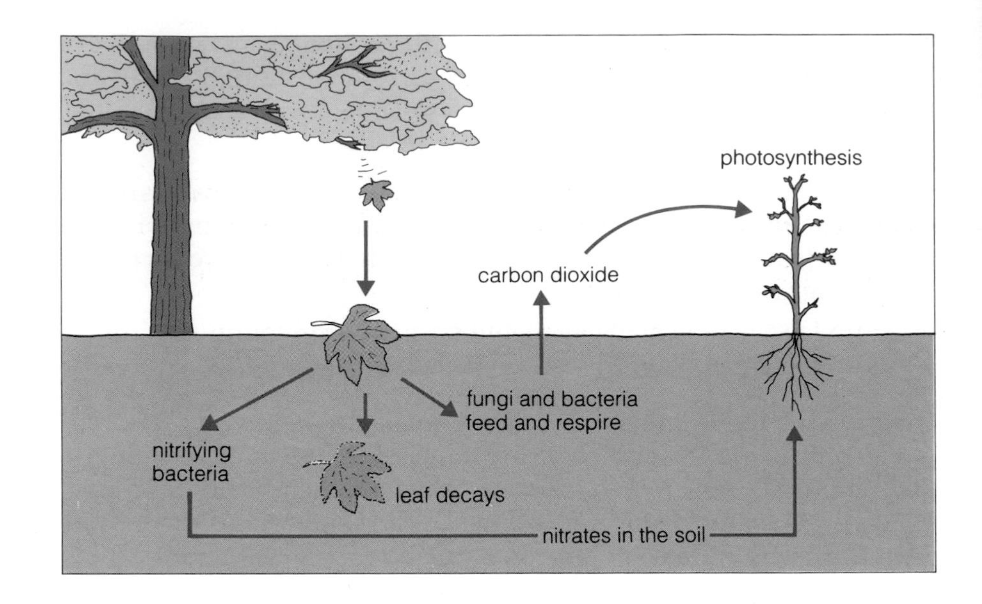

Farming

As crops grow, they gradually remove the nutrients from the soil. These have to be replaced. The traditional way of doing this was to **fertilise** the land with animal dung. Now, artificial fertilisers, which have levels of nitrate, phosphate, and potassium tailored to individual crop needs, are used to replace the nutrients in the soil and to boost crop production.

Problems can occur if too much fertiliser is put on the land. Nitrates, in particular, are very soluble. Rain can wash them off the land and into streams. High levels of nitrates in ponds and streams encourage the growth of microbes. This increases the demand for oxygen in the water. Eventually, a shortage of oxygen kills the plants, animals and fish living in the water.

Did you know?

- Bacteria in the roots of beans, peas and clover can help them 'fix' nitrogen from the air into nitrates that plants can use.
- Sometimes the environmental conditions are not suitable for bacteria and fungi to live, and so decay does not occur. This man was alive 2500 years ago. His remains were preserved in a peat bog where it is too acidic for bacteria to survive.

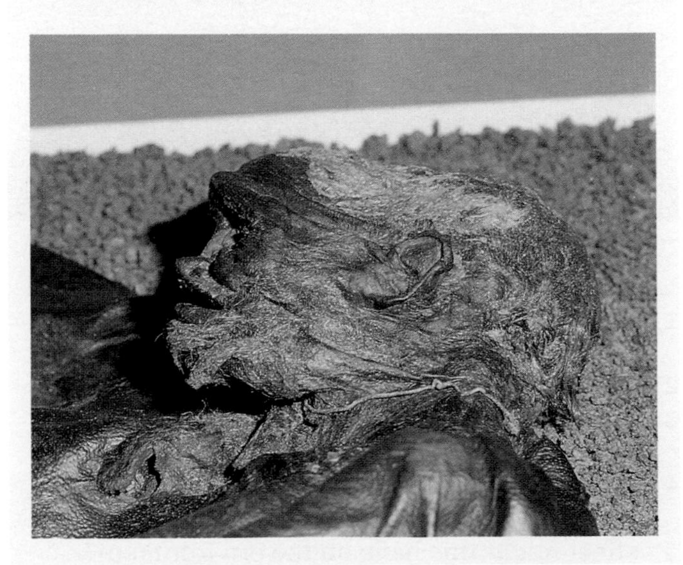

1. Why is it important that dead animals and plants decay? ▲
2. Name three nutrients in the soil which plants use to grow. ▲
3. Sort this list into things which can be broken down by bacteria and fungi, and things which can't.
   orange, tyre, wooden crate, drink can, apple, cardboard box, leather shoe
4. In your own words, describe the benefits and potential problems of using fertilisers on farms. ▲

20.2 That food's for me!

Going further 2

Being able to grow more food is great - unless something else eats it first. For every single human being on Earth, there are 50 million insects who would like their share!

Insects aren't the only thing that can reduce crop yields. Plant diseases can cause crops to wither and rot. And weeds can take nutrients from the soil which could be used by the main crop.

Farmers tackle these 'enemies' with **pesticides** (chemicals which kill *pests*).

Herbicides (weedkillers)

Unless treated, weeds will be harvested along with the main crop.

Fungicides

This damage is caused by a fungus growing in warm, moist conditions.

Insecticides

Spraying fruit trees to prevent attack by insects.

More problems

But pesticides, too, can damage the environment. Many of them are poisonous substances. Heavy rain can wash them off the land and into ponds and streams. Once they enter the food chain, fish, plants and animals in the water may die.

But it's even more far-reaching than that. Many pesticides contain chemicals which are very persistent. This means that they are very slow to break down into harmless substances. The chemical stays harmful in the bodies of animals that have eaten sprayed leaves, and in the bodies of the predators of these animals.

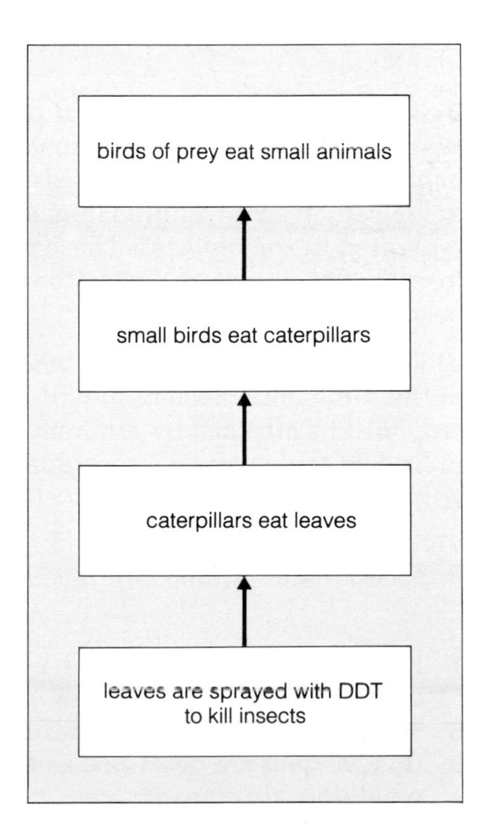

Did you know?

- DDT (short for dichlorodiphenyltrichloroethane!) was a widely used insecticide in the 1940s. In the 1960s it was found to be causing the deaths of large numbers of birds of prey.

1. Why is there a need for pesticides? ▲
2. How can a pesticide kill animals it was not meant to kill? ▲
3. In the photographs above, why is the weed spray at a low level and the insect spray high?
4. Sometimes, pesticides are sprayed onto crops from low flying aircraft. What extra risks could this bring?
5. Why should spraying be not carried out when it's windy?

20.2 Farming methods

For the enthusiast

Artificial pesticides and fertilisers increase the quantities, or **yield** of crops from an area of land. However, they can be damaging to the environment and some farmers prefer to use methods which do not rely on the use of chemicals. Foods produced by these methods are called **organic**. Farming methods which involve artificial fertilisers and pesticides are called **intensive**.

Crop rotation is a traditional way of getting the best out of the land. Each field is sown with a different crop each year. By choosing crops which have different **nutrient** needs, the soil can nourish each crop in turn. One year a field might have peas on it. Remember that peas can fix nitrogen in the soil. Next year it might be planted with potatoes, and in the third year, wheat. Then, the field might be sown with grass for grazing. The dung from the animals provides a new supply of nutrients to the land. At the end of the year, the grazing might be ploughed back into the soil, to add further nutrients.

Peas

Potatoes

Wheat

Rotation has two other advantages. Different crops are attacked by different **pests** and by different **diseases**. Changing the crop each year helps to prevent each type of pest and disease from getting established in the field.

Biological pest control Natural predators can be used to control pests. When hedges are removed to make larger fields, this removes the habitat of many insects and birds. Some of these are **predators**. Larger fields can mean fewer predators and more pests. Hedges are now being replanted in some places. These will provide habitats for birds and insects. The predators among them will help to reduce the numbers of pests which damage crops.

'Double cropping' is a system where two crops are grown side by side in the same field. Each crop will have different nutrient needs. Each crop will be attacked by different pests. Scientists are experimenting to find suitable pairs of crops. Ideally, predators living on one crop will eat the pests attacking the other crop.

These methods *reduce* the need for chemicals. For economic reasons, it is likely that farming will never be *completely* organic.

Orange predatory mites attack spider mites, a pest of bean plants.

1. In your own words describe what crop rotation is, and what the benefits are. ▲
2. What is biological pest control? ▲
3. If vegetables are described as **organically** grown, what does this mean?
4. Give an example of biological pest control.
5. **Try to find out** about intensive and organic methods for producing meat.

20.3 The need for energy

Starting off

Energy is important for everyone. Your own personal supply of energy comes from your food. But energy is also needed to *produce* that food. Energy is needed, too, for heating and for transport. Industry needs energy to obtain raw materials and to convert them into finished products.

You will know that amounts of energy are measured in **joules** (J). But one joule of energy is a very small amount. It's about the energy needed to lift this book 30 cm above your desk. So a more useful unit is the **megajoule** (MJ). This is one million joules.

In the United Kingdom, about 200 000 MJ of energy are used every year for *each person* in the country. If you had to store your personal share of that energy it would be equivalent to keeping about six tonnes of coal in your back garden! Let's have a look at some examples to show how these quantities can mount up:

It takes 21 MJ of energy to produce one loaf of bread.

It takes 90 000 MJ of energy to make a car.

It takes 600 000 MJ of energy to make a house.

It takes 8000 MJ of energy to make a refrigerator.

Forty years ago, it took only about 15 MJ of energy to make the average loaf of bread. This included the energy for harvesting the wheat, grinding the flour and baking the bread. But less artificial fertiliser was used then, so less energy was needed to manufacture fertiliser. There were more local bakeries, so the energy needed for transport was less. The bread was likely to have been unsliced, so no energy was needed for the slicing process. Plain paper would have been used for wrapping, so the energy to produce packaging would have been less.

Changing methods of production such as these mean that our overall energy needs have continued to grow.

More than 90% of the energy used in Great Britain is obtained from **coal**, **oil** and **gas**. You will remember that these are called **fossil fuels**. They were produced in the Earth's crust from the *fossilised* remains of trees, plants and sea creatures. These lived and died millions of years ago. Fuels are burnt, and the heat is used to generate electricity. There are limited amounts of fossil fuel. The more we take out, the less there is left. Fossil fuels are **non-renewable** resources. Coal and oil are also used to make soaps, detergents, plastics, and synthetic materials among many other things.

Did you know?

- The unit of energy was named after James Prescott Joule, the son of a brewer in Salford. He was interested in the relationship between mechanical energy and heat. He is said to have spent most of his honeymoon measuring the difference in water temperature between the top and bottom of a waterfall!

1. What unit is energy measured in? ▲
2. How many joules are there in: a) a kilojoule b) a megajoule?
3. Make a list of ten of the energy-using processes needed to turn wheat into a loaf of bread on the shelf of your local supermarket.

20.3 Renewable energy sources 1

Because fossil fuels will eventually run out, scientists are developing other ways of generating energy. You have already studied nuclear energy. Other methods include using the energy from the Sun, the wind, the sea and from plants. These are all **renewable** energy sources. This means that however much energy you use, the source does not run out. These renewable energy sources can reduce the damage to the environment. They don't produce pollution or nasty waste products. But they each have their own disadvantages.

Energy source	For	Against
Hydro-electric Energy Potential energy is stored in the water contained in reservoirs high in the hills. When water is released, it flows down pipes and gains kinetic energy. This is used to turn turbines and to generate electricity.	Every time it rains, there is more water to provide more energy.	Large dams have to be built and valleys may need to be flooded to provide the store of water. This may destroy wildlife habitats (including rainforests). It can only be used in wet, hilly regions.
Solar Energy Solar energy can be used in two ways: ● collector panels are used to convert the Sun's energy directly into heat energy ● panels of solar cells are used to convert the Sun's energy into electrical energy.	Whenever the Sun shines heat energy is produced. However much energy you use, more energy will be produced. Energy from collector panels is produced quite cheaply. Solar cells are made from silicon, which is an abundant element.	It only works when the Sun shines. Collector panels are only useful for producing small increases in temperature (e.g. warming water). Solar cells are still quite expensive to produce. For larger amounts of energy, large areas of solar cells are needed.
Wind Energy The Sun heats up the land and results in winds forming. The wind drives windmills or wind generators. These produce mechanical or electrical energy.	Whenever the wind blows energy is produced. However much energy you use, more energy will be produced.	It only works efficiently in windy places. For large amounts of energy very many generators covering a large area are needed.
Wave Energy The wind forms waves on the ocean. The energy in the wave is used to produce electricity by generators.	Whenever there are waves, energy is produced. However much energy you use, more energy will be produced.	Wave generators only work efficiently in exposed sea areas (e.g. around the north west coast of Scotland). For large amounts of energy, many kilometres of wave generators are needed.

1 What does renewable mean? ▲
2 What is the advantage of using renewable energy sources? ▲
3 "We depend on the Sun for most of our energy needs". Explain how this is true for a) the sources above and b) fossil fuels. ▲
4 Why does it make sense to look for alternatives to fossil fuels to provide our energy?
5 **Try to find out** more about generating energy from the wind.

Wind generators

Did you know?

- We use more electrical energy during the day than at night.
- At Dinorwic, in North Wales, water falls from a high level reservoir to produce electricity during the day. At night, spare electricity from fossil fuel power stations is used to pump water back to the high level reservoir. This provides water ready to supply the next day's energy needs.

20.3 Renewable energy sources 2

For the enthusiast

Energy source	For	Against
Tidal Energy The movement of the tides, produced by the gravitational pull of the Moon (and to a smaller extent, the Sun), is used to drive turbines to produce electricity.	Whenever there are tides, energy is produced. However much energy you use, more energy will be produced at the next tide.	It only works efficiently in areas with a large tidal range (e.g. the River Severn estuary). A large area of water must be enclosed by a barrage. The high water levels behind the barrage can alter wildlife habitats.
Geothermal Energy In some areas, the high temperatures within the Earth heat up rocks close to the surface. Cold water is pumped into these rocks and the hot water produced can be used for local heating schemes.	Whenever there are hot rocks near the Earth's surface, energy can be produced. However much energy you use, more is available.	It's only possible in areas where there are hot rocks near the Earth's surface (e.g. certain parts of Iceland and New Zealand).

Energy can also be obtained from biomass. A variety of processes can be used. Each involves using the energy stored in living plant materials. Remember that plants use the Sun's energy to convert carbon dioxide and water into sugars (photosynthesis).

A methane generator

Growing plants to burn Plants, especially fast growing trees, can be grown for fuel. **Growing plants for oil** The seeds of certain plants such as sunflowers and oil seed rape contain high levels of oil. This oil can be used as fuel as well as in foods. **Fermenting biomass** Sewage products, dung and plants can be fermented to produce methane gas. The sugar from sugar cane can be fermented to produce alcohol.	These sources *are* renewable: however much you use, you can grow some more. In Brazil, alcohol from sugar was mixed with petrol and used as a fuel for specially converted car engines.	The burning fuel *does* release carbon dioxide into the atmosphere. This tends to be only in small quantities. For large energy production, large land areas would be needed. Artificial fertilisers might have to be used. For certain schemes, a suitable climate is important. Remember that burning large quantities of **existing** forest timber can have long-term effects on the atmosphere.

1 Do tidal energy and geothermal energy depend on the Sun? ▲
2 What damage can tidal energy schemes do to the environment? ▲
3 What advantage does using biomass have over fossil fuels? ▲
4 What are likely to be the main impacts on the environment from using energy from biomass?
5 One method of making more efficient use of our energy sources would be to use less energy! **Try to find out** how a 'Save it' scheme could reduce the energy used in:
a) house heating and lighting, b) transport.

Did you know?

- One 'petroleum nut' tree can produce 50 litres of oil each year for cooking or lighting.
- In a suitable climate, a farmer putting 10% of the land to sunflowers could grow enough oil to power farm machinery.
- 1200 ha of the fast-growing 'ipilipil' tree can produce in a year the same energy as 1 million barrels of crude oil.

20.4 The need for materials

Starting off

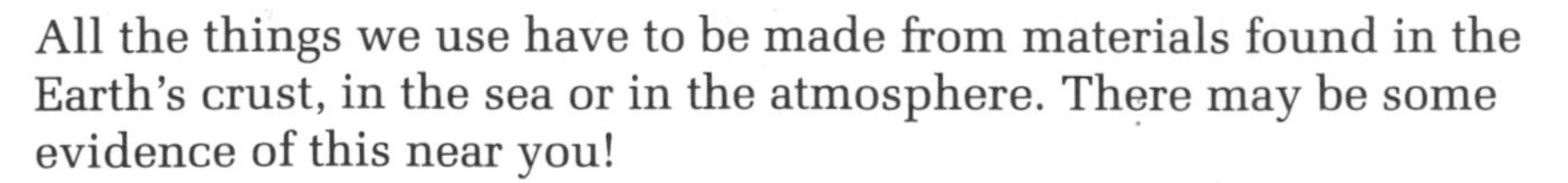

All the things we use have to be made from materials found in the Earth's crust, in the sea or in the atmosphere. There may be some evidence of this near you!

Our society is always wanting new products. This means that we have to keep on obtaining a fresh supply of raw materials. This puts pressure on the planet in two ways. The planet's materials themselves are being used up, and energy is used in obtaining them.

Some types of food would be unhygienic if not wrapped to protect them.

Packaging

However, there are ways in which the need for raw materials can be reduced. Most of the things we buy are wrapped. Sometimes the packaging is there for a very good reason. You couldn't sell a drink, for instance, without a container to put it in. And some delicate pieces of equipment have to be protected against damage. But in other cases, the packaging is just for show. The wrapping is designed to be attractive so that you are tempted to buy the product. Sensible use of materials can save valuable resources.

Recycling

Another way of making good use of materials is to recycle things when they get to the end of their useful lives. There are two ways of doing this.

Sometimes things can be re-used again quite easily. A jumble sale is a good source of re-usable objects.

In another form of recycling, the material from old objects is reprocessed to make new things. This saves on the amount of raw materials and the amount of energy needed. Paper, aluminium, glass, plastics and iron and steel are all things that can be reprocessed.

The main difficulty with recycling is collecting and separating the old materials. It doesn't make sense, for instance, to use lots of petrol to take one bottle to the bottle bank! You would use more of the Earth's resources in oil than you would save in glass.

A milk bottle must be the ideal re-usable container. It takes 5.5MJ of energy to make it. It can be washed and refilled many times.

It can take 30g of aluminium and 7MJ of energy to produce a can from new materials. It takes one old can and less than 0.5MJ of energy to make a can from recycled materials.

Did you know?

- 44% of Britain's steel is made from recycled materials.
- Less than 5% of household waste is recycled.
- In some areas, trials are being made to separate household waste before it is collected. You would have one container for clean paper. You would have other containers for steel, glass, aluminium and plastics.

1. Make a list of any industrial developments near you which are concerned with obtaining raw materials.
2. List some examples of wasteful packaging that you have seen. For each example, explain how you would package it better.
3. What are the advantages of recycling? ▲
4. Why doesn't it make sense to drive to a collection point to take only small quantities of cans, bottles or paper? ▲
5. **Try to find out** why more drinks manufacturers don't use 'returnable and refillable' bottles.

20.4 Processing materials: pollution

Going further 1

It's not only energy production which produces waste materials — so does the manufacturing industry.

Some industrial developments look ugly. Some factories cause noise pollution and keep their neighbours awake at night. But those things don't effect the Earth as a whole.

More wide-reaching are the wastes which many industrial processes produce. These wastes can be in three forms:

- Solids which have to be dumped
- Liquids which are pumped into rivers or to sewage treatment plants
- Gases which are released into the atmosphere.

Wastes can be classified into two broad groups.

- **Persistent** wastes. These chemicals stay the same over a period of time. They do not break down into simpler substances.
- **Biodegradable** wastes. These are broken down over a period of time by microbes. Methane gas is produced as the wastes decay.

Scientists have developed biodegradable plastics. Some supermarkets are now using plastic bags and containers which are biodegradable. This is an advantage when the article is not designed to be recycled.

Persistent detergents form a foam on ponds and streams. They harm wildlife. Many modern detergents are now biodegradable.

Solid wastes are usually tipped and buried. Often the waste is tipped in former quarries. These are called 'landfill' sites. After a tip has been left to settle for many years, it can be built on. As the amount of waste increases there may be a shortage of tipping sites in the future.

Biodegradable detergent.

Did you know?

- Once biodegradable plastics have decayed, the material is not available for recycling.

Biodegradable waste rots down well, but produces methane. In large quantities this can cause problems. It can be dangerous if it is still present when houses have been built on the tip.

Poisonous (**toxic**) chemicals are often persistent. They can be a potential hazard over long periods of time.

After rain, chemicals from a tip can leak away into ponds and rivers. Persistent toxic chemicals can get into the food chain. These can kill animals and humans.

1. How could you protect yourself against noise pollution?
2. What does 'biodegradable' mean? ▲
3. Try to give two advantages and two disadvantages of using biodegradable plastics? ▲
4. For what reasons is landfill tipping not ideal? ▲

20.4 Processing liquid wastes

Going further 2

Liquid wastes from industry are sometimes treated on site to remove impurities. The liquid is then pumped into rivers or the sea. Other companies pump their effluent directly to sewage processing plants. There, it is treated along with household sewage.

In both methods, there is a risk that not all the impurities are removed. Dangers occur when toxic chemicals are absorbed into the **ecosystem**.

Human sewage is biodegradable, but decay takes time, and certain harmful bacteria can live in sea water for extended periods.

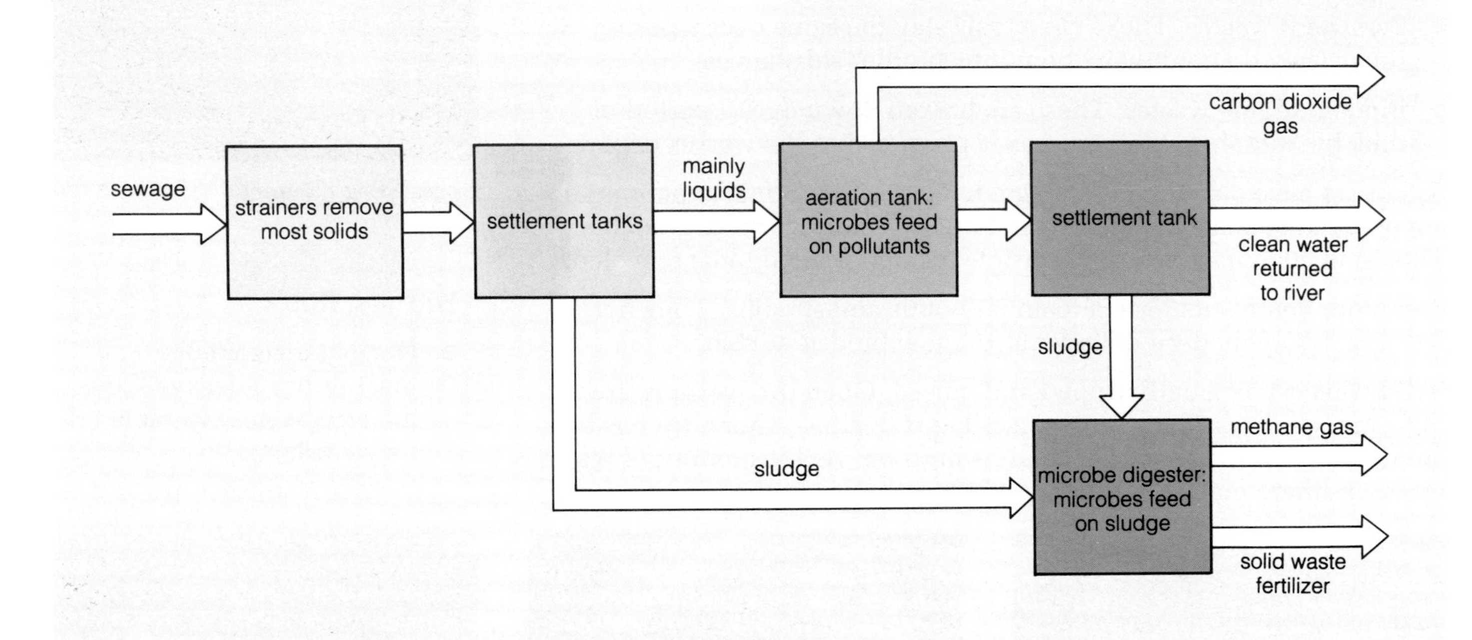

The water returned to rivers or the sea after treatment may still be polluted. Shellfish near sewage outfalls are often inedible. There is a need to improve water quality in the future. In the 'Root Zone' system, effluent is fed into a bed of reeds before being released. Microbes living in the roots break down toxic wastes into simple non-toxic compounds. Although a large reed bed is needed, the process is economical to operate.

1. Draw a flow diagram of the sewage treatment process. Include a 'root zone' stage at the end of the process. ▲
2. The sludge digestion stage produces methane and solids. What could each of these be used for?
3. Some sewage works dump digested sludge in the sea. What effects could this have?
4. Why might shellfish near a sewage outfall be inedible?

Did you know?

- The Llanwern steelworks in South Wales has an 18 hectare reed bed for treating liquid effluents produced during coke making.

20.4 The greenhouse effect

For the enthusiast

Pollution is not a new problem. It has been growing since the start of the Industrial Revolution about 200 hundred years ago. But over the last 50 years it has been accelerating. It is only in the past few years that scientists have realised just what damage is being done to the environment.

The greenhouse effect

Without its atmosphere the Earth would be much colder. Its average temperature would be below freezing point. Gas molecules in the atmosphere act rather like the glass in a greenhouse. The Earth's average temperature is raised to about 22°C.

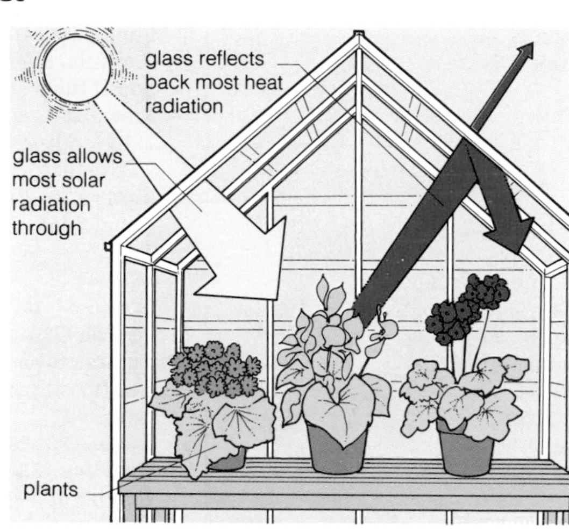

In a greenhouse, sunlight comes in through the glass. Much of the light energy is absorbed by the soil and is changed into heat energy. The glass prevents the heat from escaping to the outside. Inside the greenhouse it becomes warmer than outside.

Sunlight travels down through the atmosphere. Much of the light energy is absorbed by the Earth and is changed to heat. Molecules of water vapour, carbon dioxide and other gases in the atmosphere act like the glass in the greenhouse. They prevent all the heat escaping to space.

Many industrial processes produce gases which increase the **greenhouse effect**. These include carbon dioxide, methane, nitrogen oxides and chlorofluorocarbons (or **CFCs**). Some gases increase the effect more than others. One molecule of methane gas has 30 times the 'greenhouse effect' of one carbon dioxide molecule.

Chlorofluorocarbons (CFCs) are used in aerosol cans, fridges, freezers, and to make plastic foams. One CFC molecule has 17 000 times the greenhouse effect of a carbon dioxide molecule.

The amount of carbon dioxide in the atmosphere is increasing. Burning rain forests and burning coal, oil and gas in power stations are just two of the causes. Some scientists believe that this could lead to a permanent increase in the temperature of the Earth.

Because plants take in carbon dioxide during photosynthesis, they are important in keeping levels of carbon dioxide in the air stable. Photosynthesis happens quickly in hot, wet conditions. Tropical rain forests have lots of leaf area. They use up large quantities of carbon dioxide from the atmosphere.

1 Why would saving energy help to reduce the greenhouse effect? ▲
2 Make a list of gases which contribute to the greenhouse ▲
3 What results do you think that global warming could have?
4 Give two reasons why burning rainforests could affect global warming.

Did you know?

- Doubling the carbon dioxide level in the air would increase the Earth's average temperature by around 4°C.
- Many scientists think that CFCs also damage the ozone layer. This is a gas layer about 50 kilometres above the Earth's surface which helps to protect us from the Sun's harmful ultra-violet radiation.

Index

Acknowledgements

The publishers wish to thank the following for permission to reproduce photographs.

Ancient Art and Architecture Collection, p.78; Ashmolean Museum, Oxford, p.75; Continental Wine Experts, p.49 (top left); Corrosion and Materials Consultants Ltd, p.50 (middle); Flour Advisory Bureau, p.49 (top right); Equinox, p.12; Mary Evans Picture Library, p.22 (left), p.30 (lower), p.44; p.55, Genesis Space Photo Library, p.11 (top and middle); W.L. Gore Ltd, p.40 (bottom right); Robert Harding Picture Library, p.10, p.33 (bottom); Michael Holford, p.40 (top right, bottom left); Holt Studios International, p.79 (middle), p.80 (top), p.83; Chris Honeywell/OUP, p. 23 (all), p.26 (top right), p.155 (all), p.41 (all), p.48 (bottom), p.50 (top), p.51 (all), p.58 (right), p.63 (top), p.64, p.65 (top), p.81 (middle), p.84 (top); The Hutchinson Library, p.7, p.22 (right), p.37 (all); ICI plc, p.49 (bottom); Rob Judges, p.33 (top); Kos Photos, p.69 (right); Andrew Lambert, p.64 (bottom); Jerry Mason, p.58 (left and middle), p.61 (middle), p.63 (bottom); Nasa, p.4; National Dairy Council, p.26 (top left); National Medical Slide Bank, p.24, p.29 (centre left and far right), p.31, p.32 (bottom); Natural History Photographic Agency, p.72, p.73 (top), p.77 (all), p.79 (right and left), p.85 (bottom); Oxford Scientific Films, p.71 (left), p.73 (middle bottom); Philip Harris Education, p.71 (bottom); Philips, p.68; Linda Proud, p.26 (bottom), p.50 (bottom), p.71 (right), p.81 (left and top right), p.84 (middle), p.85 (top); Ann Ronan Picture Library, p.28, p.40 (middle); Saint Bartholomew's Hospital, London: Department of Medical Illustration, p.32; Science Photo Library, p.1 (bottom left), p.2, p.6, p.11 (bottom), p.15, p.16, p.17, p.18, p.19, p.20, p.29 (top, far left, centre right), p.32 (top), p.34, p.42, p.56, p.65 (bottom), p.80 (bottom), p.82, p.86; Sharp Electronics, p.61 (left and right); Frank Spooner Pictures, p.14, p.33 (centre); Zanussi Ltd, p.69 (middle), p.81 (lower right); Zefa Picture Library, p.1 (top).

Illustrations by Nick Hawken, Nic Lipscombe, Joe Little, Fiona Plummer and Rodney Sutton.

Produced by V.A.P. Publishing Services, Kidlington, Oxfordshire.